Advanced Tools and Models to Improve River Basin Management in Europe in the Context of Climate Change (AquaTerra)

Advanced Tools and Models to Improve River Basin Management in Europe in the Context of Climate Change (AquaTerra)

Edited by
Michael Finkel, Johannes Barth and Peter Grathwohl

IWA Publishing
London • New York

Published by **IWA Publishing**
Alliance House
12 Caxton Street
London SW1H 0QS, UK
Telephone: +44 (0)20 7654 5500
Fax: +44 (0)20 7654 5555
Email: publications@iwap.co.uk
Web: www.iwapublishing.com

First published 2010

Typeset in India by OKS Prepress Services.

British Library Cataloguing in Publication Data
A CIP catalogue record for this book is available from the British Library

Library of Congress Cataloging-in-Publication Data
A catalog record for this book is available from the Library of Congress

ISBN 10: 1843393727
ISBN 13: 9781843393726

Contents

Authors

Hans-Jørgen Albrechtsen, Danish Technical University, Dept. of Environmental Engineering, Lyngby, Denmark (hja@env.dtu.dk)

Erik Ansink, VU University Amsterdam, Institute for Environmental Studies, Amsterdam, The Netherlands (erikansink@gmail.com)

D. Baes, University of Liege, Gembloux Agro-Bio Tech (baes.d@fsagx.ac.be)

Jean-Michel Baltassat, BRGM, Orléans, France (jm.baltassat@brgm.fr)

Olivier Banton, Université d'Avignon et des pays de Vaucluse (UAPV), Avignon, France (olivier.banton@univ-avignon.fr)

Nicole Baran, BRGM, Service EAU, Orléans, France (n.baran@brgm.fr)

Damià Barceló, CSIC, Institute of Environmental Assessment and Water Research, Department of Environmental Chemistry, Barcelona, Spain and Catalan Institute for Water Research (ICRA), Girona, Spain (damia.barcelo@idaea.csic.es)

Anthony Bardos, Cyber Sales Ltd, Reading, UK (anthony.bardos@cybersales.co.uk)

Paul Bardos, r3 Environmental Technology Ltd, Dept. Soil Science, Reading, UK (paul@r3environmental.co.uk)

Johannes A.C. Barth, Geo-Center of Northern Bavaria, Applied Geology Group, Erlangen, Germany (barth@geol.uni-erlangen.de)

Jordi Batlle-Aguilar, Ecole Polytechnique Fédérale de Lausanne (EPFL), School of Architecture, Civil and Environmental Engineering, Ecological Engineering Laboratory, Lausanne, Switzerland (jordi.batlle@epfl.ch)

Alberto Bellin, University of Trento, Department of Civil and Environmental Engineering, Trento, Italy (alberto.bellin@ing.unitn.it)

Enrico Bertuzzo, Ecole Polytechinque Federale, Laboratory of Ecohydrology, Facultè ENAC, Lausanne, Switzerland (enrico.bertuzzo@epfl.ch)

Stephen Blenkinsop, Newcastle University, School of Civil Engineering and Geosciences, Newcastle, UK (s.blenkinsop@ncl.ac.uk)

Martino Boni, Università di Padova, Department IMAGE, Padova, Italy (martino.boni@ciclistiamari.it)

Gianluca Botter, Università di Padova, Department IMAGE, Padova, Italy (gianluca.botter@unipd.it)

Madjit Bouzit, BRGM, Montpellier, France (m.bouzit@brgm.fr)

Isabella Bovolo, Newcastle University, School of Civil Engineering and Geosciences, Newcastle, UK (Isabella.Bovolo@ncl.ac.uk)

Agnès Brenot, BRGM – Service EAU, Orléans, France (a.brenot@brgm.fr)

Jos Brils, Deltares, Utrecht, The Netherlands (jos.brils@deltares.nl)

Hans Peter Broers, Deltares, Unit Soil and Groundwater Systems, Utrecht, The Netherlands (hanspeter.broers@deltares.nl)

Serge Brouyère, Universite de Liege, Departement ArGEnCo, Secteur Geő, Liège, Belgium (Serge.Brouyere@ulg.ac.be)

Claudius Bürger, Hydrogeology Group, University of Tübingen, Tübingen, Germany (claudius.buerger@uni-tuebingen.de)

Aidan Burton, Newcastle University, School of Civil Engineering and Geosciences, Newcastle, UK (Aidan.Burton@ncl.ac.uk)

Tony Chapman, wca environment ltd, Faringdon, UK (Tony.Chapman@wca-environment.com)

Dominique Darmendrail, BRGM, Direction Générale, Orléans, France (d.darmendrail@brgm.fr)

Sylvia Dautrebande, (Dautrebande.sylvia@skynet.be)

Michael Finkel, University of Tuebingen, Center for Applied Geosciences, Tuebingen, Germany (michael.finkel@uni-tuebingen.de)

Hayley Fowler, Newcastle University, School of Civil Engineering and Geosciences, Newcastle, UK (h.j.fowler@ncl.ac.uk)

Francis Garrido, BRGM, EPI/ECO, Orléans, France (f.garrido@brgm.fr)

Martin Gerzabek, University of Natural Resources and Applied Life Sciences, Vienna, Austria (martin.gerzabek@boku.ac.at)

Sylvain Gigleux, Agence National pour la Gestion des Déchets Radioactifs – ANDRA, Châtenay-Malabry, France (Sylvain.Gigleux@andra.fr)

Jean-François Girard, BRGM – Natural Risks, Orléans, France (jf.girard@brgm.fr)

Tilmann Gocht, Environmental Mineralogy & Chemistry Group, University of Tübingen, Tübingen, Germany (tilman.gocht@uni-tuebingen.de)

Pascal Goderniaux, University of Liège, Group of Hydrogeology and Environmental Geology – Aquapôle, Liège, Belgium and Funds for Scientific Research – FNRS, Brussels, Belgium (Pascal.Goderniaux@ulg.ac.be)

Peter Grathwohl, University of Tübingen, Center of Applied Geoscience, Tübingen, Germany (peter.grathwohl@uni-tuebingen.de)

Nina Graveline, BRGM, Département Eau, Montpellier, France (n.graveline@brgm.fr)

Alexis Gutierrez, BRGM, Water Division, Orléans, France (a.gutierrez@brgm.fr)

Cécile Herivaux, BRGM, Water Department, Orléans, France (c.herivaux@brgm.fr)

Jean-Pierre Hermand, Universite Libre de Bruxelles, Brussels, Belgium (jean-pierre.hermand@ulb.ac.be)

Pei-Chi Hsu, Hamburg University of Applied Science (HAW), Hazardous Substances and Ecotoxicology, Hamburg, Germany (pei-chi.hsu@haw-hamburg.de)

G.S. Janniche Gry Sander Janniche, Technical University of Denmark, Department of Environmental Engineering, Lyngby, Denmark (gsf@env.dtu.dk)

Kevin C. Jones, Lancaster Environment Centre, Centre for Chemicals Management, Lancaster University, Lancaster, UK (k.c.jones@lancaster.ac.uk)

Frédéric Jouin, BRGM, Orléans, France (jouin.frdrc@gmail.com)

Jan Joziasse, Deltares, Soil and Groundwater Systems, Utrecht, The Netherlands (jan.joziasse@deltares.nl)

Edda Kalbus, Helmholtz-Centre for Environmental Research – UFZ, Department of Hydrogeology, Leipzig, Germany (edda.kalbus@ufz.de)

Gerard Th. Klaver, Deltares, Geo Environmental Research Laboratory, Utrecht, The Netherlands (Gerard.Klaver@deltares.nl)

Thomas Klinka, FOOTWAYS, Orléans, France (t.klinka@footways.eu)

Bart Koelmans, Wageningen University, Aquatic Ecology and Water Quality Management Group, Wageningen, The Netherlands (bart.koelmans@wur.nl)

Olaf Kolditz, Helmholtz Centre for Environmental Research – UFZ, Department of Environmental Informatics, Leipzig, Germany (olaf.kolditz@ufz.de)

Ruben Kretzschmar, Swiss Federal Insitute of Technology, Department of Environmental Sciences, Zurich, Switzerland (ruben.kretzschmar@env.ethz.ch)

Georg J. Lair, University of Natural Resources and Applied Life Sciences, Vienna, Austria (georg.lair@boku.ac.at)

Alette Langenhoff, Deltares, Subsurface and Groundwater Systems, Utrecht, The Netherlands (alette.langenhoff@deltares.nl)

Ingo Lobe, Helmholtz Centre for Environmental Research – UFZ, Department River Ecology, Leipzig, Germany (ingo.lobe@ufz.de)

Bruno Majone, University of Trento, Department of Civil and Environmental Engineering, Trento, Italy (bruno.majone@ing.unitn.it)

Marco Marani, Università di Padova, Department IMAGE, Padova, Italy (marco.marani@unipd.it)

Franz-Josef Maringer, BOKU – University of Natural Resources and Applied Life Science, Department of Forest- and Soil Sciences – PLUS, Vienna, Austria (franz-josef.maringer@boku.ac.at)

Annemieke Marsman, Deltares, Utrecht, The Netherlands (annemieke.marsman@deltares.nl)

Sandra Meijer, The REACH Centre Limited, Lancaster Environment Centre, Lancaster University, Lancaster, UK (s.meijer@thereachcentre.com)

Corinne Merly, BRGM, Orléans, France (c.merly@brgm.fr)

Peter Middeldorp, A&G Milieutechniek bv, Waalwijk, The Netherlands (Peter.Middeldorp@aengbedrijven.nl)

Romain Millot, BRGM, Metrology, Monitoring, Analysis Division – MMA/ISO, Orléans, France (r.millot@brgm.fr)

Claudia Moeckel, Lancaster Environment Centre, Lancaster University, Lancaster, UK (c.moeckel@lancaster.ac.uk)

Christophe Mouvet, BRGM, Orléans, France (c.mouvet@brgm.fr)

Philippe Négrel, BRGM – Service EAU, Orléans, France (p.negrel@brgm.fr)

Philippe Orban, University of Liège, Belgium (HGULg), Liège, Belgium (p.orban@ulg.ac.be)

Anibal Perez-Garcia, University of Tübingen, Hydrogeology Group, Tübingen, Germany (anibal-jose.perez-garcia@student.uni-tuebingen.de)

Emmanuelle Petelet-Giraud, BRGM – Service EAU, Orléans, France (e.petelet@brgm.fr)

Mira Petrovic, CSIC, Institute of Environmental Assessment and Water Research, Department of Environmental Chemistry, Barcelona, Spain and Catalan Institution for Research and Advance Studies (ICREA), Barcelona, Spain (mira.petrovic@idaea.csic.es)

Anton Poot, DHV, Amersfoort, The Netherlands (anton.poot@dhv.com)

Frido Reinstorf, Helmholtz Centre for Environmental Research – UFZ, Department of Hydrogeology, Leipzig, Germany and Magdeburg-Stendal University of Applied Sciences, Dept. Water and Waste Management Magdeburg, Germany (frido.reinstorf@ufz.de)

Huub H.M. Rijnaarts, Wageningen University, Division Environmental Technology, Wageningen, The Netherlands (huub.rijnaarts@wur.nl)

Marc Rijnveld, TNO, Delft, The Netherlands (marc.rijnveld@tno.nl)

Andrea Rinaldo, Ecole Polytechinque Federale, Laboratory of Ecohydrology, Facultè – ENAC, Lausanne, Switzerland and Università di Padova, Department IMAGE, Padova, Italy (rinaldo@idra.unipd.it)

Stéphane Roy, BRGM, Service EPI, Orléans, France (s.roy@brgm.fr)

Mario Schirmer, Eawag – Swiss Federal Institute of Aquatic Science and Technology, Department Water Resources and Drinking Water, Dübendorf, Switzerland (Mario.Schirmer@eawag.ch)

Christian Schmidt, Helmholtz Centre for Environmental Research – UFZ, Department of Hydrogeology, Leipzig, Germany (christian.schmidt@ufz.de)

Diana M.E. Slijkerman, IMARES, Department of Environment, Den Helder, The Netherlands (Diana.Slijkerman@wur.nl)

A.F.L. Slob Adriaan F.L. Slob, TNO, Delft, The Netherlands (adriaan.slob@tno.nl)

J. Slobodnik Jaroslav Slobodnik, Environmental Institute, Kos, Slovak Republic (slobodnik@ei.sk)

D. Steidle Dietmar Steidle, University of Tübingen, Center for Applied Geosciences, Tübingen, Germany (dietmar.steidle@uni-tuebingen.de)

Philippe Stollsteiner, BRGM/EAU/ADM, Orléans, France (p.stollsteiner@brgm.fr)

Pierre Strosser, Ecologic Institute, Berlin, Germany

Claire van der Wielen, Institut Scientifique de Service Public (ISSeP), Liège, Belgium (cl.vanderwielen@issep.be)

C.A.M. (Kees) van Gestel, VU University Amsterdam, Animal Ecology, Amsterdam, The Netherlands (kees.van.gestel@falw.vu.nl)

Bertil van Os, Ministry of Education, Culture and Science, Cultural Heritage Agency, Amersfoort, The Netherlands (B.van.Os@cultureelerfgoed.nl)

Michelle van Vliet, Wageningen University, Earth System Science and Climate Change group, Wageningen, The Netherlands (michelle.vanvliet@wur.nl)

Michel Verbanck, Universite Libre de Bruxelles, Dept. Water Pollution Control, Brussels, Belgium (mikeverb@ulb.ac.be)

Jos P.M. Vink, Deltares, Group Sediments and river basins, Utrecht, The Netherlands (Jos.Vink@deltares.nl)

Ate Visser, Deltares, Delft, The Netherlands (Ate.Visser@deltares.nl)

Tomas Vogel, Czech Technical University, Faculty of Civil Engineering, Prague, Czech Republic (vogel@fsv.cvut.cz)

Wolf von Tümpling, Helmholtz Centre for Environmental Research – UFZ, Department of River Ecology, Water Analytics & Chemometrics, Leipzig, Germany (wolf.vontuempling@ufz.de)

Arnaud Wuilleumier, BRGM, Service EAU, Orléans, France (a.wuilleumier@brgm.fr)

Mauricio Zambrano-Bigiarini, University of Trento, Department of Civil and Environmental Engineering, Trento, Italy (mauricio.zambrano@ing.unitn.it)

Stefano Zanetti, Università di Padova, Department IMAGE, Padova, Italy (zanetti@idra.unipd.it)

Chapter 1

Introduction

P. Grathwohl, M. Finkel and J. A. C. Barth

With the onset of mining, man started to distribute elements on the earth's surface which were until then mostly in the earth's crust. During many centuries of mining operations – sometimes more that 1000 years – environmental compartments such as river sediments and subsequently floodplain soils became enriched with heavy metals and other pollutants. In addition, with the increase of the population after the medieval era the increased burning of biomass and coal caused the deposition of soot, chars and associated organic compounds such as the carcinogenic polycyclic aromatic hydrocarbons (PAHs) on soils. Sites with coal based industries (ore smelters, glass production, charcoal piles) produced severe local soil pollution. After the onset of the industrial revolution emission of pyrogenic compounds into the atmosphere and soils increased by a factor of approximately 100, mostly connected due to increased energy demand covered by fossil fuel (coal and oil).

In the 20th century, manufactured chemicals for many industrial, agricultural and household purposes were released into the environment in large quantities. Some of the most prominent groups include halogenated compounds that are for instance used as pesticides (DDT, lindane, PCP), flame retardants (PCBs later PBDE), solvents (TCE, PCE, etc.). These manmade compounds are inherently stable in most environments and therefore occur now at a global scale and throughout the food web. Meanwhile, there is no doubt that these manmade compounds produced a chemical footprint on the whole planet, which is an often overseen but important aspect of global change.

Through comprehensive sampling and monitoring campaigns in diverse European river basins and by using newly developed field and laboratory methods including passive samplers, soil column and microbiological experiments, AquaTerra could show that many manmade chemical compounds can be found today ubiquitously, in all compartments, in air, in soil and sediments, in surface waters as well in groundwater.

Fate and transport of these compounds in the environment is highly dynamic and numerous physical, chemical as well as biological processes govern the pathway of pollutants in the environment. This includes complex input functions into soils and surface waters, migration in soil, groundwater, surface waters and sediments as well as redistribution processes from soil and water into the atmosphere.

Following the "chemocycle" of pollutants, AquaTerra made integral efforts to identify and parameterize the biological, chemical and physical processes relevant for the prediction of the long-term fate and impact of man-made pollutants in the environment.

Despite environmental regulations such as the Stockholm Convention REF on persistent organic pollutants (regulating the so-called "dirty dozen"), new compounds are continuously designed, produced with the risk that they will eventually enter the environment.

Many novel measurement and analytical techniques developed in AquaTerra allow to detect an increasing number of compounds at increasing levels of detail.

Not only do scientists need to understand how systems might function particularly with respect to transfer of pollutants, but also do societies need to react to environmental loads of pollutants, their transfer and accumulation on an international basis. The latter becomes increasingly important if large scales are considered.

AquaTerra's work was also based on strong stakeholder involvement, regarding knowledge production as a social process in order to present scientific results in a manner that allow use for a wide a range of potential end users as possible.

Top soils serve as filter for deposited pollutants and protect groundwater bodies efficiently against diffuse pollution in many European areas. Resilience may apply today but accumulation in soils and sediments results in increasing concentration of many compounds that may compromise resilience of soils and associated ecosystems. Concentrations of many compounds are still relatively low and often found at trace levels (except at hot spots e.g. contaminated sites, which often act as primary sources), but there is a trend of not just increasing concentrations but also in the number and complexity of compounds appearing in the environments. Whether these constantly changing compound cocktails cause any toxicological effects on man or biota is as of yet unclear. New scientific themes therefore include questions whether low concentrations of xenobiotic compounds matter at all and how fast organic compounds are degraded or diluted in the environment. In short, how much pollutants can the system take before a "real" damage occurs? Diffuse pollution is not easily reversible and the question arises how permanent a ubiquitous chemical footprint is.

To answer these vital questions, AquaTerra made a significant step forward in the development of process-based numerical models that are capable of delivering predictions of likely trends and environmental changes to be expected in the near and distant future. These models can be used to generate hydrologic scenarios based on climate models and to simulate pollutant distribution and turnover rates from decades to millennia.

This book gives an overview of the activities and achievements of AquaTerra. More detailed information are provided on the project website (http:www.eu-aquaterra.de).

Chapter 2

HYDRO – Climate and water cycle at the basin scale

H. J. Fowler, A. Gutierrez, M. Marani, S. Blenkinsop, A. Burton, P. Stollsteiner, S. Brouyère, P. Goderniaux, M. van Vliet, S. Zanetti, J. F. Girard, S. Dautrebande, D. Baes, C. Mouvet, N. Baran, J. M. Baltassat, C. I. Bovolo, A. Bellin, B. Majone, M. Zambrano-Bigiarini, M. Boni, F. Jouin, T. Klinka, G. S. Janniche, H. J. Albrechtsen, A. Wuilleumier, S. Gigleux and O. Banton

KEY OBJECTIVES

The key objectives of HYDRO were to provide:

- an improved understanding of the mechanisms and impact of global change, and in particular climate change, on the water cycle
- relevant historic climatic data for assessments of the impact of climatic variability on water quality and availability, as well as key biogeochemical processes in soils and sediments and how these may change under perturbations of climate and land-use
- innovative methods in generating climate scenarios; modelling of rainfall properties e.g. inter-annual variability, intensity and seasonality at a range of space/time scales
- downscaling of climate model output to generate climate change scenarios
- geological and hydrogeological data to improve understanding of hydrogeological/ hydrodynamic phenomenon and contribute to the understanding of pollutant transfer
- a better characterisation of hydrology and hydrodynamics in the saturated and unsaturated zones to improve the understanding of pollutant transfers
- assessments of water inputs and outputs to the Brévilles catchment at Montreuil-sur-Epte, France (Figure 2.1).

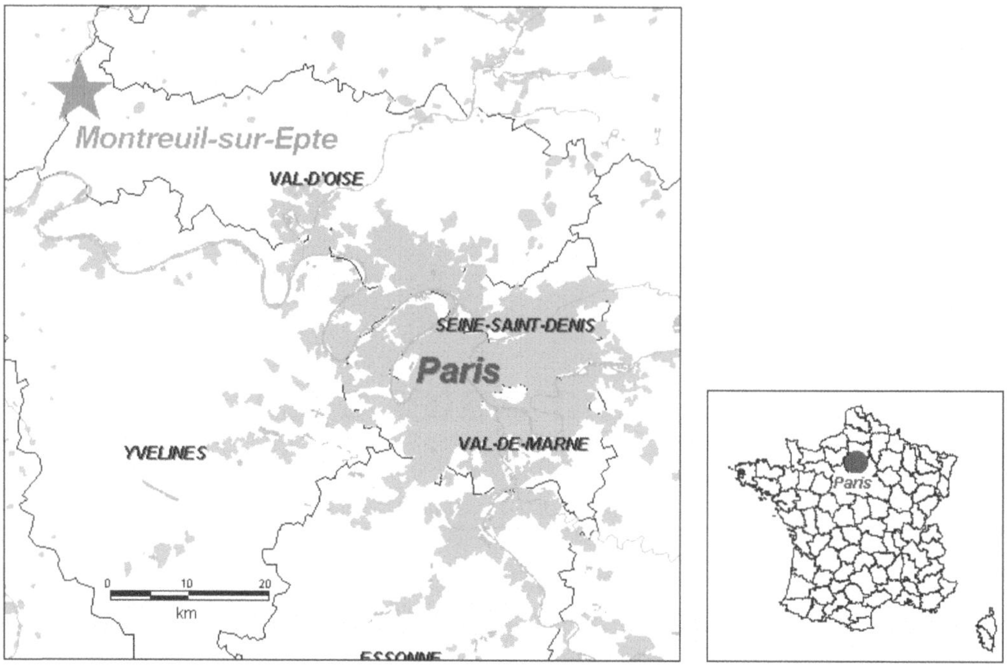

Figure 2.1 Location of the site of Montreuil-on-Epte.

KEY ACHIEVEMENTS

Framework for probabilistic climate change scenarios

The climate change downscaling method developed within HYDRO is a framework for the construction of probabilistic climate change scenarios for the assessment of impacts on hydrological and hydro-geological systems at regional-to-catchment scales (Fowler and Wilby, 2007; Fowler *et al.*, 2007a). A key aim of the method is to produce probabilistic projections of future change by using multiple climate model realisations of future climate, enabling the quantification of uncertainties in future projections and allowing the estimation of the likelihood of the risk (with respect to set thresholds).

Water availability predictions

Water Scarcity: Regional climate model projections for a precipitation-derived drought severity index (DSI6) for the end of the century (Figure 2.2) show that (i) there will be more long summer droughts (>6 months) in southern Europe associated with a >30% decrease in rainfall amount and (ii) there will be more frequent short term (3–6 month) droughts in northern Europe (Blenkinsop and Fowler, 2007a,b). In northern Europe, winter rainfall is projected to increase, thus inhibiting long-term droughts whilst summer rainfall is projected to decrease, thus increasing the probability of short-term summer droughts. For the Dommel catchment (Meuse Basin, Netherlands), future change in drought is uncertain whilst for the Gallego (Ebro Basin, Spain), climate model projections suggest that droughts will be more frequent and longer. Estimates of change are likely to be conservative as the additional effects and feedbacks associated with likely temperature increases were not included in the analysis.

Extreme rainfall events: Changes to short and long duration extreme rainfall were examined for 1, 2, 5 and 10 day events (Fowler *et al.*, 2007b). Climate models simulated increases in extreme rainfall events across Europe, even in areas which may experience a reduction in mean rainfall (Figure 2.3; from Fowler *et al.*, 2007b). Longer extreme events are projected to increase with larger increases in the biggest events compared with more moderate events, particularly in northern Europe. In climates affected by convective extremes, e.g., the Mediterranean region, projections are uncertain as climate models cannot reproduce these processes adequately and therefore underestimate extreme rainfall events in these regions.

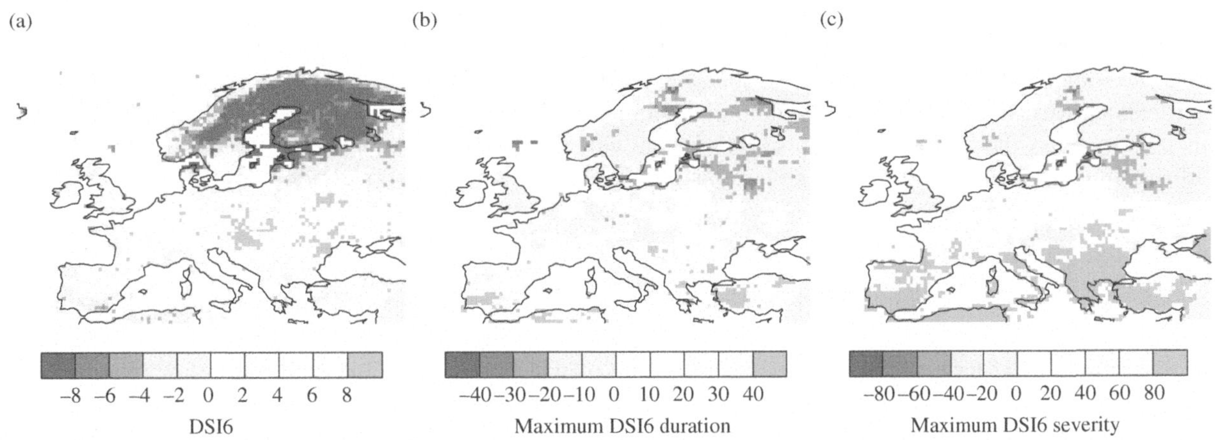

Figure 2.2 Projected change in: (a) the frequency of DSI6 events (>6 months duration), (b) the maximum DSI6 duration in months, and (c) the maximum DSI6 severity, for an experiment with the HadRM3P regional climate model provided by the PRUDENCE project[1]. All changes are expressed for the period 2071–2100 relative to 1961–1990.

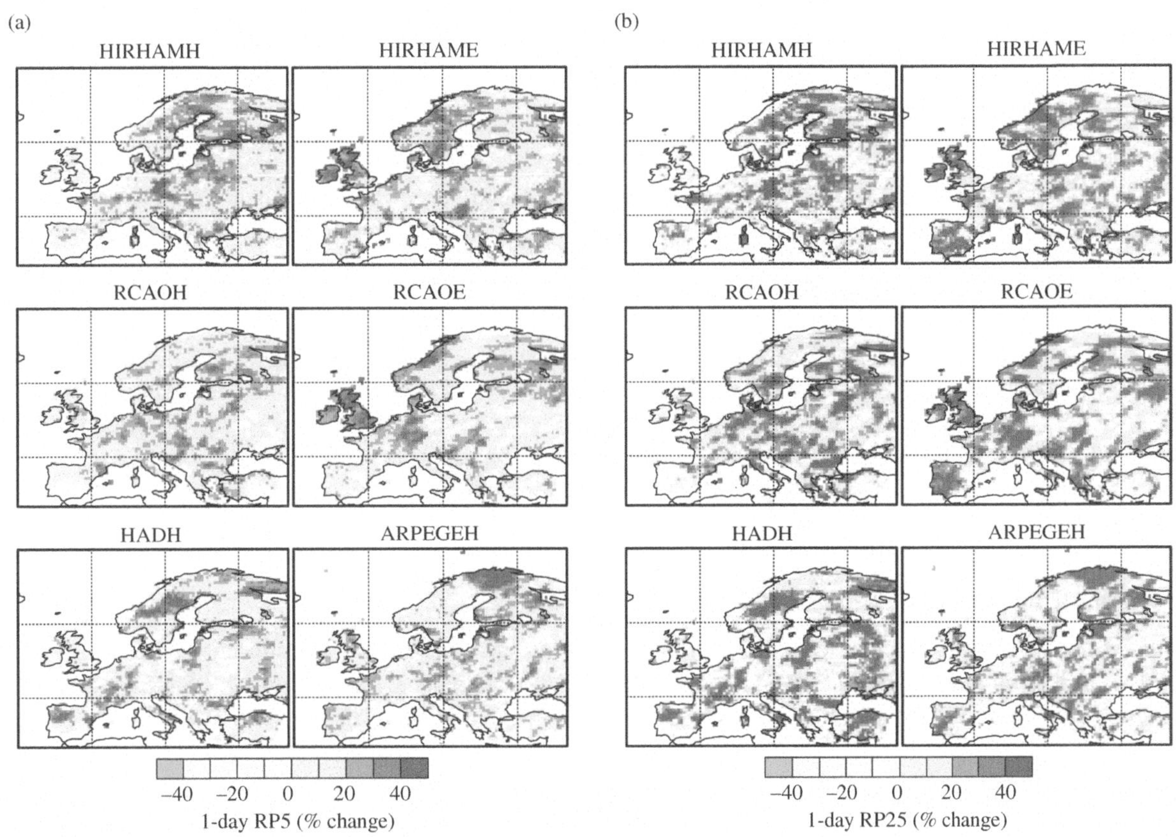

Figure 2.3 Estimates of percentage change in return values for (a) 1 day, 5 year events and (b) 1 day, 25 year events for all European grid cells as derived from an ensemble of experiments provided by the PRUDENCE project. All changes are expressed for the period 2071–2100 relative to 1961–1990.

Spatial downscaling

The level of detail that must be incorporated in space-time rainfall fields used to force hydrological models to provide reliable estimates of the associated flood hydrographs was investigated. By using extensively validated geomorphological models of hydrologic response applied to a wide variety of basins, up-scaling procedures were developed to

[1] Regional climate model experiments provided by PRUDENCE can be downloaded from http://prudence.dmi.dk/

decrease the spatial detail with which rainfall variability is described, starting from extremely refined grid resolution (0.1×0.1 km^2) up to uniform rainfall. For catchments with areas up to 8000 km^2, the main source of variability in flood response was found to be the average rainfall instantaneously injected into the catchment. Therefore, even for quite large basins (areas up to several thousand km^2), rainfall forcing can be specified using a coarse resolution (of the order of 10 km^2). This shows that results from global or regional climate models may not require spatial downscaling as the rainfall structure at scales of ~1 km does not have an appreciable effect on the flood hydrograph (Nicotina *et al.*, 2008).

Temporal downscaling

New theoretical results were achieved, connecting the variance of temporal and spatial rainfall to the scale of observation under the assumption of second-order stationarity (Marani, 2003; 2005). In particular, it was shown that the theoretically-derived form of the variance as a function of scale is incompatible with power-law relations usually assumed for the second-order statistical moment within multi-scaling approaches to rainfall modelling. Validations making use of a dataset representing a wide variety of resolutions, climates and observation instruments show a good agreement between observations and the theoretically-predicted forms of the variance as a function of aggregation. The proposed theoretical framework has then been used to construct general relations for statistical rainfall downscaling based on the estimation of the hourly variance on the basis of daily observations (Marani and Zanetti, 2007). The estimated hourly variance has been used in a Bartlett-Lewis stochastic model to generate hourly synthetic time series preserving observed statistics. The resulting stochastic downscaling procedure was then applied to rainfall observations for the Dommel, Brenta and Ebro catchments (Zanetti and Marani, in preparation).

Stochastic 'weather' generation

Weather generator methods have been developed, allowing the concurrent simulation of precipitation and other meteorological parameters (e.g., relative humidity, temperature, wind speed, etc.) useful in continuous hydrological modelling. Two approaches have been developed: the first is based on a multivariate auto regressive scheme, allowing the generation of correlated sequences of hourly precipitation, wind direction and speed, pressure, tidal storm surge, daily minimum and maximum temperature, daily minimum and maximum relative humidity, incoming total radiation. A second, stochastic weather generator tool produces multiple simulations of future weather variables by applyling change factors derived from regional climate model control and future scenarios. The software is available as freeware for the UK (Kilsby *et al.*, 2007) and has been extended to model the Dommel (Van Vliet, in preparation), Brévilles (Burton *et al.*, 2010) and Geer (Blenkinsop *et al.*, in preparation; Goderniaux *et al.*, in preparation) catchments within AquaTerra. For the Brévilles and Geer this incorporated a novel transient, or time-evolving change, rather than the conventional "time-slice" approach. This provided scenarios which can be used on timescales of relevance to local policy makers and planners.

Non-stationarity and climate change scenarios

Additional work has been achieved developing new optimisation approaches for the calibration of a suite of rainfall models used in hydrological modelling (Burton *et al.*, 2008) and developing a non-stationary stochastic rainfall model for catchments where non-stationary rainfall occurrence processes are prevalent (Burton *et al.*, submitted).

Establishment of a monitoring network

A monitoring network has been set-up for the Brévilles catchment, which follows the former site characterization undertaken in the framework of the FP5 PEGASE project. This network is presented in Figure 2.4. This monitoring network includes:

Observation wells: the number of access points to the water table has increased from 8 to 22. New drillings (project deliverable H2.1bis, see www.eu-aquaterra.de) have brought information regarding: geology (depth of geological layers); lithology (nature and heterogeneity of aquifer rocks); and hydrodynamics: (permeability of saturated and unsaturated zone) (project deliverable H2.5).

Springs: A new gauging station was built in 2005 to monitor the Brévilles spring. Several measurements under different hydraulic conditions were performed to obtain relationships between level and discharge in the channel 200 m downstream. Other springs have been added to the monitoring network. They are manually gauged twice a year (project deliverable H2.2).

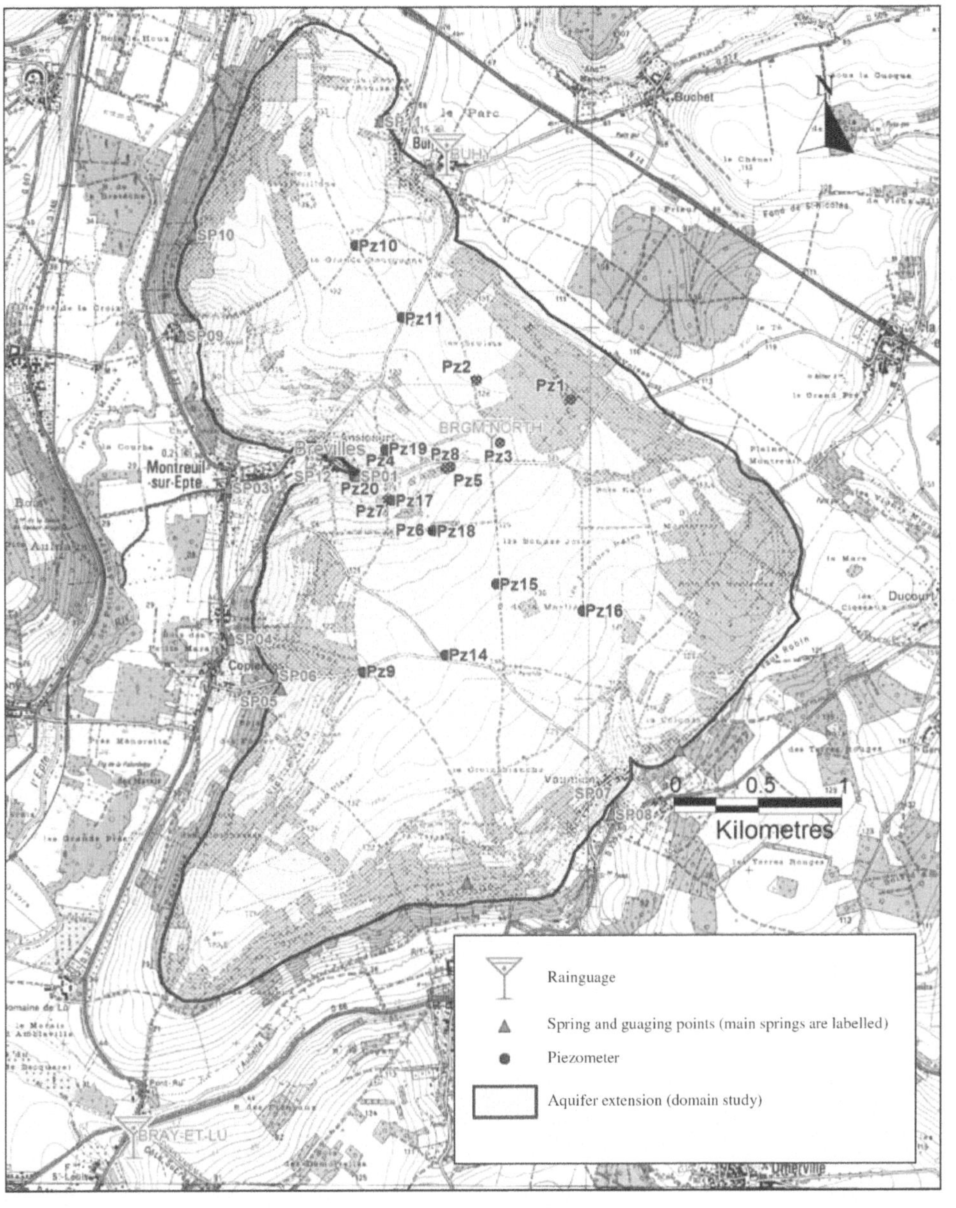

Figure 2.4 The Brévilles monitoring network in 2008.

Raingauges: the French meteorological office has two climatic stations in the area. BRGM is managing one catchment raingauge. A reference rainfall series at a daily time step from 1988 to present and at a monthly time-step from 1950 was established (project deliverable H2.6).

Pumping tests in drilled wells: Pumping tests were performed in most of the 22 observation wells to monitor water levels during and after phases of pumping at different discharges. The tests have confirmed higher permeability in the upper part of the aquifer than in the lower part.

Catchment Investigations

Geophysics: To improve the hydro-geological knowledge, geophysical methods were applied to characterize the aquifer. Subsurface structures were studied using 2D electrical resistivity tomography (ERT) along 6 profiles across the basin. The ERT results clarified the locations of the main faults and found additional faults and the presence of a marly layer at an elevation of around 120 m (Figure 2.5). More than 60 magnetic resonance soundings (MRS) were performed between 2000 and 2005. MRS is a new technique in hydrogeology and its application to Brévilles catchment is part of the method validation (Girard *et al.*, submitted). This gives an indication of the free water content in the geological formations and may be calibrated to an accuracy of ~2 m on the water depth. This interpretation helped in the drawing of a piezometric map and the definition of a realistic infiltration area, ~4.4 km^2 (project deliverable H2.4). Both the geophysical and drilling results enabled the elaboration of the geological structure which is the basis of the mathematical model developed within the COMPUTE module (project deliverable C2.8) aiming to establish pesticide content prediction in Brévilles water (project deliverable T2.10).

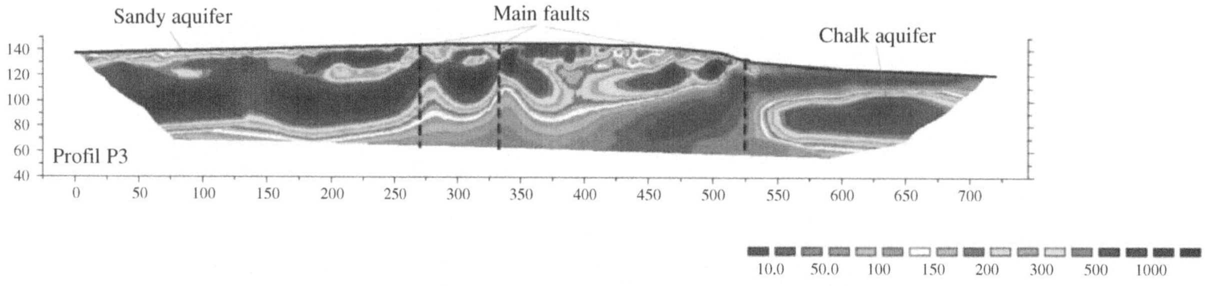

Figure 2.5 Example of ERT interpretation (section across eastern faults).

Tracer tests: Injections of a known volume of various solutes into several piezometers and following their restitution at the spring proved the existence of spatial heterogeneity. Two piezometers located at less than 200 m upstream from the spring were tested in 2003 with two different dyes. Only one dye appeared, 7 days later at the spring. After 300 days, only 22% of the mass of dye was recovered. This atypical concentration evolution was attributed to tracer capturing in the underground system, probably close to the injection point. In November 2005 additional experiments were performed using a new method, Finite Volume Point Dilution (FVPDM) (Brouyère *et al.*, 2008). Four wells were tested using four different tracers. Among the 4 tracers directly injected into the aquifer through piezometers, only 2 appeared at the spring and at the weir, 200m downstream, confirming the 2003 experiment. The tracer injected at the top of the aquifer appeared first, confirming that flow is faster in the upper part of the aquifer. The breakthrough curves show that several velocities are involved, corresponding to different underground pathways (Figure 2.6). A preliminary interpretation estimated effective porosity as ~20% in the upper part of the saturated zone and 10 to 14% in the lower part (project deliverable H2.3).

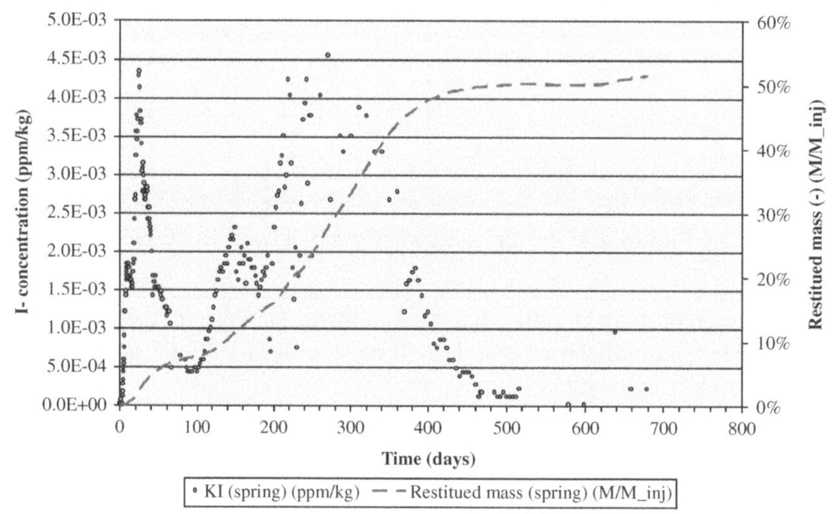

Figure 2.6 Concentration of tracer (Iodide injected at Pz17b) at the Brévilles spring.

IMPLICATIONS AND APPLICATION OF RESULTS

HYDRO results have enabled recommendations to be made on how the climate system may behave under climate change, useful for both climate change modellers whose remit includes the modelling of climate and prediction of climate change, and river basin managers who are responsible for managing water quantity issues. Additionally, HYDRO has extensively studied the small catchment of the Brévilles spring which, despite its apparent simplicity, is an extremely rich and varied case study. The system is small with significant recharge each year; underground flow is however very slow and complex. This strong inertia explains why the environmental measures taken 8 years ago at the entrance of the soil/sediment/groundwater system are not yet seen at its exit.

CONTRIBUTIONS AND POLICY RECOMMENDATIONS FOR RIVER BASIN MANAGEMENT

HYDRO has shown that climate change scenarios indicate that there will be an increase in extreme events related to changing rainfall patterns. This means an increase in short-term drought events in northern Europe, and longer term droughts in southern Europe, an increase in intense rainfall everywhere, and longer duration rainfall events in northern Europe. River basin managers must be able to absorb these changes into systems that are not necessarily well-adapted to the extremes that we already see today. Since the direction of potential changes are reasonably well understood but the probable magnitude of the change is not so clear then river basin managers should concentrate on the reduction of current vulnerabilities to extreme events; i.e., increasing the robustness of our river basins and water supply systems to flood and drought events while bearing in mind that these types of events are likely to increase in frequency and magnitude in the future.

HYDRO has also shown the importance of the hydro-geological context on underground flow and transport processes (Baran *et al.*, 2007). Diffuse pollution observed at the spring results from an underestimation of aquifer vulnerability, inappropriate agricultural policy and lack of knowledge of pesticide residence times and toxicity. Taking into account the hydrogeological characteristics of the Brévilles catchment would have allowed the prediction of the spring contamination and appropriate regulation could have been taken. Diffuse pollution has been neglected during the past 50 years, specifically, the French procedure for water well protection is only now beginning to systematically consider the impact of diffuse pollution in the setting up of protection zones and findings from the Brévilles study could become an incentive to do so in the future.

HYDRO has provided catchment scale evidence that water transfers can be so slow that any environmental measure aiming at reducing the contamination of drinking water abstraction may take a very long time to take effect (Gutierrez and Baran, 2009). A side effect of this is the probable negative impact on the motivation of stakeholders to apply or respect carefully the regulation/recommendations. Therefore, river basin management should more fully consider groundwater protection and contamination prevention. This can be done by conducting programs such as HYDRO in water supply catchments (springs and boreholes) and also by integrating agricultural practices into water resources management.

Locally/nationally

Projections of change in drought and extreme rainfall event frequency and magnitude have been produced at the local scale. These can help managers to develop management strategies to mitigate some of the impacts of climate change. The simulations generated by the weather generator (primarily of rainfall, temperature and potential evapotranspiration) are very useful input data for hydrological and hydro-geological models, which will in their turn enable the prediction of the consequences of heavy rainfall on processes such as floods and contaminant transport.

Because pesticide and nitrate concentrations exceeded water quality limits, the district council of the 350 inhabitant village of Montreuil-sur-Epte was forced to abandon a cheap water resource and subscribe to a more expensive water supply system. Consequently, the price of water rose and the rural population is longing for their spring, which is considered by locals to have a better taste than the present distributed water. The research study carried out at Montreuil-sur-Epte, and particularly field works conducted within HYDRO2, opened local minds to the vulnerability of water resources and maintains hope of a return to pristine water. Results of monitoring show however that this return will probably take decades.

In a European context

Mitigation – Droughts: In order to implement river basin management measures to minimise the effects of the increase of droughts expected in the future, it is necessary to take into account how much extra unused water resource is currently present in the hydrological system (amount of water with respect to population water demand). Solutions are likely to be found at the regional scale. *Mitigation – Extreme events*: Increases in extreme rainfall events have to be taken into account by river basin managers. The management measure should be location specific. Moreover, extreme rainfall events must be managed by also taking into account droughts. For example the construction of water reservoirs may enable the channelling of rainfall from extreme rainfall events and simultaneously allow the creation of water resources in case of droughts. *The probabilistic framework for climate change scenarios* can be applied worldwide.

Any one model simulation of future climate may represent only one of many possible future climate states. Therefore, a large number of climate model simulations must be undertaken to assess the uncertainty and estimate the most likely future climate. It is therefore recommended that predictions of climate change should *use multiple different models* which may be compared in their ability to model historical climate and so their possible skill in predicting future climate be assessed.

As a result of the Water Framework directive, Atrazine is being banned from European countries. However, the effect of environmental measures depends on groundwater flow. The positive impacts of this European directive on groundwater quality will take many years before being observed in the Brévilles catchment which is not an exceptionally slow system. Groundwater systems with strong inertia exist elsewhere in France and the rest of the world and are of particular interest for water resources management as they buffer seasonal variations and offer drought resistant supplies. However, in case of contamination they offer a very long remediation time. HYDRO has shown that long-term monitoring studies are needed on such aquifers: eight years of monitoring has not established an adequate water balance for the Brévilles catchment.

OPEN & NEW QUESTIONS FROM HYDRO

HYDRO has demonstrated the need for an investigation into the validity of climate model outputs and downscaling methods as a tool for planning adaptation measures under climate change. There may be other methods that can be used other than the classic 'climate change impact study' approach. These simpler methods need to be validated and tested to see if we can improve the robustness of systems to climate change without needing detailed probabilistic information on risk. New questions include (i) the ability of regional climate models to reproduce extreme weather events necessary for planning climate change adaptation measures, (ii) how to assess the ability of climate models to simulate observed climate, (iii) how many climate models should be included in a multi-model ensemble to produce robust predictions of future changes.

At a local scale, HYDRO has opened up new interrogations regarding the functioning of the Brévilles catchment. Chosen for its apparent simplicity and small size, the Brévilles spring and its catchment is not yet fully understood. The drilling of wells and geophysical surveys showed a very heterogeneous unsaturated zone and a faulted geological structure; the monitoring of water level and spring discharge unveiled a particularly slow system and the pumping and tracer test revealed vertical variations of aquifer permeability with depth. Heterogeneities are evidenced but their organisation in space and their impact on pesticide transport are still poorly depicted and understood. The study undertaken is extremely important as it demonstrates how long contamination transfer may take and how urgent it is to take prevention and remediation measures.

REFERENCES

Baran, N., Mouvet, C., Négrel, Ph. (2007) Hydrodynamic and geochemical constraints on pesticide concentrations in the groundwater of an agricultural catchment (Brévilles, France). *Environmental Pollution*, **148**, 729–738.

Blenkinsop, S. and Fowler, H.J. (2007a) Changes in drought characteristics for Europe projected by the PRUDENCE regional climate models. *International Journal of Climatology*, **27**(12), 1595–1610.

Blenkinsop, S. and Fowler, H.J. (2007b) Changes in drought frequency and severity over the British Isles projected by the PRUDENCE regional climate models. *Journal of Hydrology*, **342**, 50–71.

Blenkinsop, S., Fowler, H.J., Harpham, C., Burton, A., Goderniaux, P., and Brouyère, S, in preparation. Modelling transient climate change with a stochastic weather generator: Projected temperature changes for the Geer catchment, Belgium.

Brouyere S., Batlle-Aguilar, J., Goderniaux P. (2008) A new tracer technique for monitoring groundwater fluxes: the Finite Volume Point Dilution Method. *Journal of Contaminant Hydrology*, **95**, 121–140.

Burton, A., Fowler, H.J., Blenkinsop, S., and Kilsby, C.G. (2010) Downscaling transient climate change using a Neyman-Scott Rectangular Pulses stochastic rainfall model. *Journal of Hydrology*, **381**, 18–32.

Burton, A., Fowler, H.J., Kilsby, C.G. and O'Connell, P.E., submitted. The Non-Stationary Activation of Raincells model for spatial-temporal simulation of non-stationary rainfall occurrence and amounts. Water Resources Research.

Burton, A., Kilsby, C.G., Fowler, H.J., Cowpertwait, P.S.P. and O'Connell, P.E. (2008) RainSim: A spatial temporal stochastic rainfall modelling system. *Environmental Modelling and Software*, **23**, 1356–1369.

Fowler, H.J. and Wilby, R.L. (2007) Beyond the downscaling comparison study. *International Journal of Climatology*, **27**(12), 1543–1545.

Fowler, H.J., Blenkinsop, S. and Tebaldi, C. (2007a) Linking climate change modelling to impacts studies: recent advances in downscaling techniques for hydrological modelling. *International Journal of Climatology*, **27**(12), 1547–1578.

Fowler, H.J., Ekström, M., Blenkinsop, S. and Smith, A.P. (2007b) Estimating change in extreme European precipitation using a multi-model ensemble. *Journal of Geophysical Research – Atmospheres*, **112**, D18104, doi:10.1029/2007JD008619.

Girard, J-F., Baltassat J.M., Gutierrez A., Gigleux S., Legchenko A., Mathieu F., Miehé J-M., submitted. Contribution of the joint use of MRS and ERT to hydrogeological modelling: case study of the Brévilles spring basin (France), *Journal of Hydrology*.

Goderniaux P, Brouyère S., Blenkinsop S., Burton A., Fowler H.J., Dassargues A, in preparation. Climate change impact on groundwater reserves using stochastic transient climate change scenarios.

Gutierrez A., Baran N. (2009) Long-term transfer of diffuse pollution at catchment scale: respective roles of soil, and the unsaturated and saturated zones (Brévilles, France). *Journal of Hydrology*, **369**, 381–391. doi:10.1016/j.jhydrol.2009.02905.

Kilsby, C.G., Jones, P.D., Burton, A., Ford, A.C., Fowler, H.J., Harpham, C., James, P., Smith, A. and Wilby, R.L., 2007. A daily weather generator for use in climate change studies. *Environmental Modelling and Software*, **22**, 1705–1719.

Marani, M. (2003) On the correlation structure of point rainfall, *Water Resour. Res.*, vol. **39**(5), doi:10.1029/2002WR001456.

Marani, M. (2005) Non-power-law scale properties of rainfall in space and time, *Water Resour. Res.*, **41**, W08413, doi:10.1029/2004WR003822.

Marani, M., Zanetti, S. (2007) Downscaling rainfall temporal variability, *Water Resour. Res.*, **43**, W09415, doi:10.1029/2006WR005505.

Nicótina, L., Alessi Celegon, E., Rinaldo, A. and Marani, M. (2008) On the impact of rainfall patterns on the hydrologic response, *Water Resour. Res.*, **44**, W12401, doi:10.1029/2007WR006654.

Van Vliet, M., Blenkinsop, S., Burton, A., Fowler, H.J., Harpham, C., Kilsby, C.G., in preparation. A multi-model ensemble of downscaled climate change scenarios for the Dommel catchment, Netherlands.

Zanetti, S., Marani, M., in preparation. Non-scaling vs. multifractal methods: a comparative appraisal with a view to rainfall downscaling.

Chapter 3

BIOGEOCHEM – Biogeochemical processes in soils and sediments

A. Langenhoff, R. Kretzschmar, F. Garrido and T. Gocht

KEY OBJECTIVES

- Identification and quantification of the key biogeochemical processes controlling the fate and transport of pollutants in soils and sediments;
- Identification of the key components (e.g., colloids, geosorbents, micro organisms);
- Impact of global change on the filter and transport functions of soils and on water quality; Improved understanding of the properties and the functional role of soils (Figure 3.1).

KEY ACHIEVEMENTS

Particles – and colloid-facilitated transport

We developed a mechanistic understanding on how colloid-facilitated transport affects the mobilization and transport behaviour of different classes of contaminants in temporary flooded soils. Microcosm experiments revealed that de novo formation of metallic copper and metal sulphide colloids (Figure 3.2) dominated the mobility of chalcophile metal contaminants including cadmium and lead during transient soil reduction (Weber *et al.*, 2009).

In contrast to chalcophile metals, arsenic was not associated with colloids in the reduced soil but was mobilized by reductive dissolution of Fe(III) hydroxides and accumulated around plant roots (Voegelin *et al.*, 2007).

By employing flow-through column experiments and zero-tension undisturbed field lysimeters it was demonstrated that the shift from oxic to anoxic conditions in response to flood events resulted in a marked release of organic colloids, polycyclic aromatic hydrocarbons and redox sensitive inorganic compounds (Figure 3.3). The prevailing redox conditions are controlled by the availability of the terminal electron acceptors that remain in the soil, mainly iron and manganese oxides/hydroxides.

Because colloid sampling in reduced soils is difficult to perform without creating oxidation artefacts, our research involved the successful development of novel tools for anoxic sample preparation and characterization by electron microscopy and synchrotron-based x-ray fine structure spectroscopy.

The laboratory findings corroborate the results obtained by lysimeter studies conducted at two locations within the Elbe catchments. Flood events were found to enhance the export of organic colloids, PAH, and redox sensitive elements including inorganic contaminants like arsenic. High molecular weight PAHs were thereby mobilized with colloid/particle fraction >0.7 μm. The key environmental factor for the release seems to be the overall duration of the inundation. The longer the flood event, the more exhaustive are the microbial mediated redox processes, which are

 Advanced Tools and Models to Improve River Basin Management in Europe in the Context of Climate Change (AquaTerra) by Michael Finkel, Johannes Barth and Peter Grathwohl ISBN: 9781843393726. Published by IWA Publishing, London, UK.

mainly responsible for the mobilization. These results confirm the findings of others, which state that climatic-driven single events are important for contaminant dynamics and mobility in soils (Totsche *et al.*, 2007). The results obtained may be used to further understand (im)mobilisation processes at other sites effected by periodic or episodic variations of water saturation, e.g., natural and constructed wetlands, fens and bogs. Periods of draught are the main climatic driver for microbially mediated redox changes at such sites, and for the formation of nanoparticulate organo-mineral aggregates, which might than act as contaminant carriers.

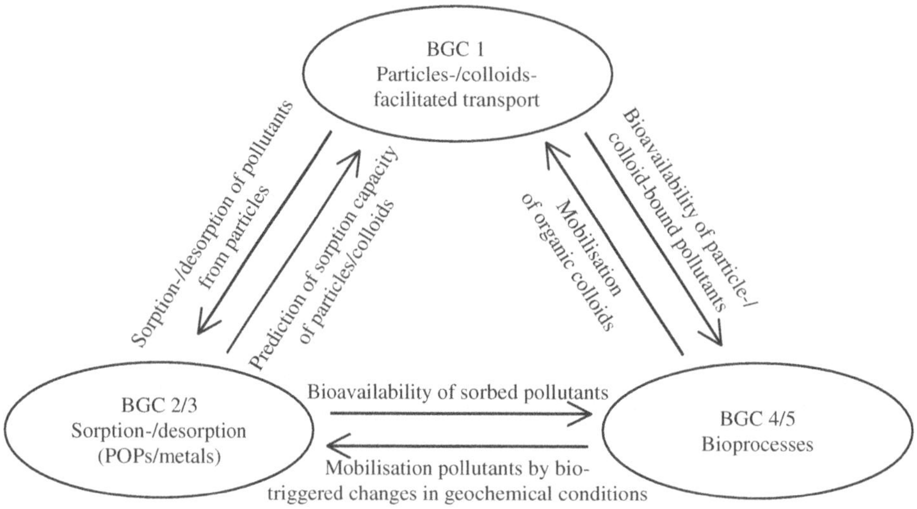

Figure 3.1 Possible relationships/ interactions between major biogeochemical processes and related work packages BGC1 to BGC5.

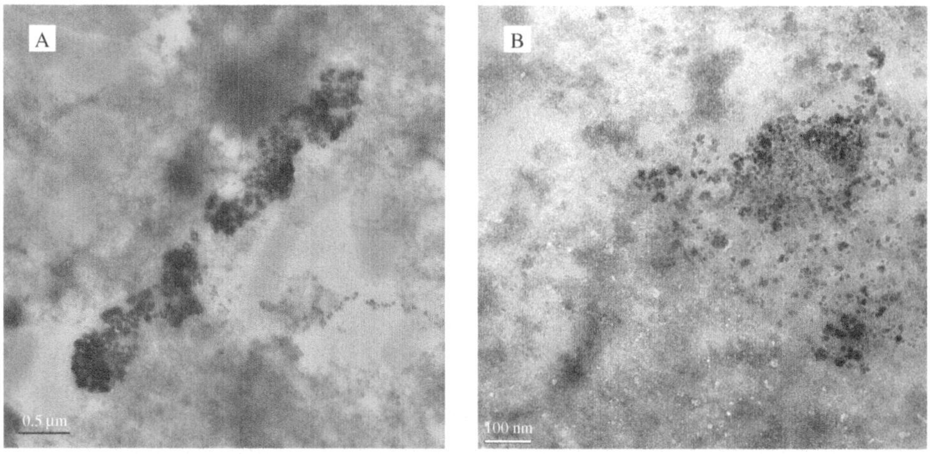

Figure 3.2 Transmission Electron Microscopy (TEM) images of (a) bacteria-associated and (b) freely dispersed copper sulphide colloids formed de novo during soil flooding in microcosm experiments. Both entities were found to carry substantial amounts of Cd and Pb dominating their transport behaviour.

Sorption processes; organic contaminants

Sorption to soils and sediments is a key process that determines the fate of contaminants in the environment. Understanding sorption processes is crucial to determine the mobility as well as the bioavailability of pollutants. For organic contaminants, the amount and type of organic matter in soils and sediments is the most important parameter that controls sorption (Abelmann *et al.*, 2005; Endo *et al.*, 2009a). Comprehensive laboratory sorption studies were conducted considering a wide variety of pollutant properties as well as different characteristics of geosorbents (Endo *et al.*, 2008; Navarro *et al.*, 2009). For all of the studied compounds the sorption capacities were highest on carbonaceous reference materials such as coal and coke. Detailed petrographic studies of samples collected from field sites among the different basins revealed the abundance of such carbonaceous particles as the result of incomplete combustion of organic materials e.g., coal, wood, and oil (Figure 3.4). Sorption capacities of selected probe compounds correlate well with the amount of these distinct fractions of organic matter in the geosorbents.

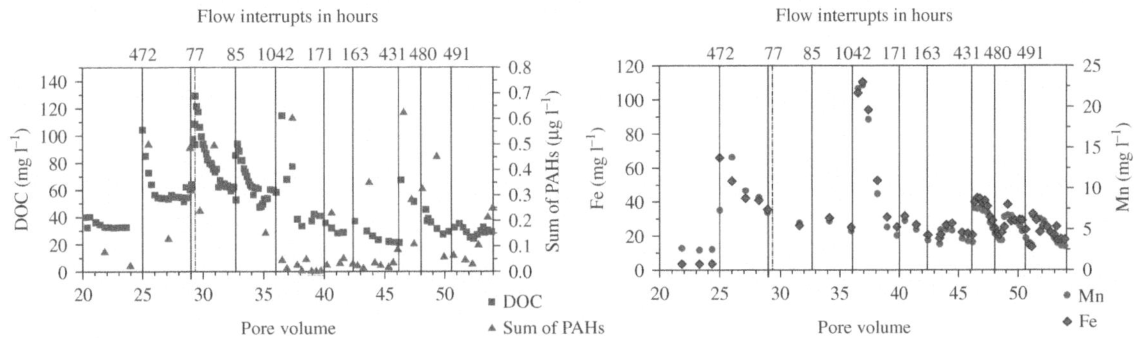

Figure 3.3 Release of DOC, PAHs, Fe and Mn before and after flow interrupts.

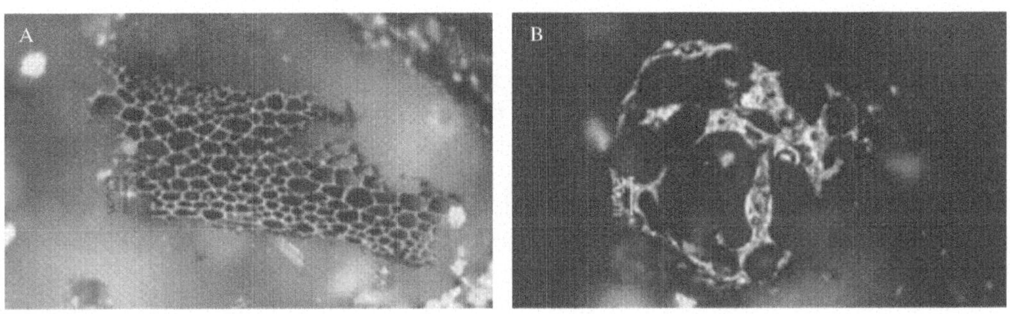

Figure 3.4 Carbonaceous particles isolated from field sites at the Bréville catchment; Combustion residue of wood, charcoal (A); Combustion residue of coal, char (B).

Besides these detailed laboratory sorption studies a model for the rapid prediction of sorption parameters such as soil-water or soil-air distribution coefficients was validated for polar and non-polar organic compounds (Endo *et al.*, 2009b). Examples of the results obtained with the model are shown in Figure 3.5. The model is based on a multiple regression approach relating particular physicochemical properties that determines molecular interactions with each other (poly-parameter linear free energy relationships). Calculated soil-water distribution coefficients using a limited number of parameters were compared with experimentally derived ones and they are in good agreement. This indicates the predictive power of the used method (Figure 3.5).

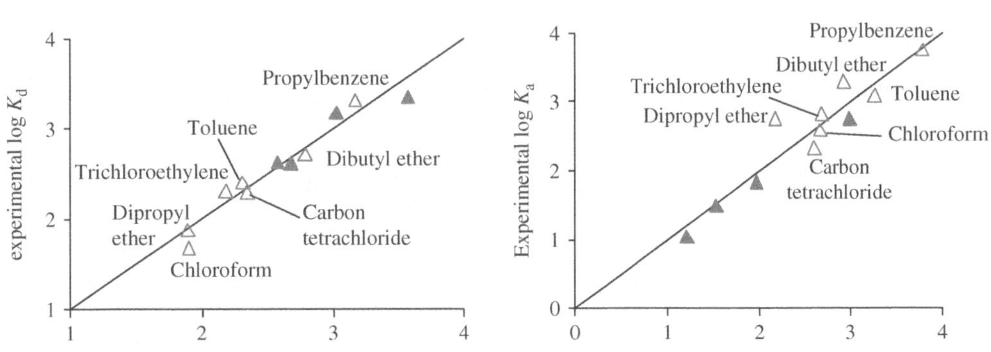

Figure 3.5 Correlation between experimental and calculated log K_d (soil-water distribution coefficient, left-hand side) or log K_a (soil-air distribution coefficient, right-hand side). The filled symbols show the data points for linear and cyclic alkanes. The solid lines represent the 1:1 line.

Release kinetics of pollutants were studied based on static as well as dynamic column tests to evaluate the filter functions of soils and sediments with respect to organic pollutants. Filter capacities are highest at the top of the soil profiles where the organic carbon content is highest as well. Once pollutants such as pesticides arrive at deeper parts of the unsaturated zone their mobility is higher due to limited sorption. This is in particular the case in limestone areas such as the Brévilles catchments.

Sorption processes; inorganic contaminants

An experimental method that is applicable to various soil systems has been used to measure the effect of four global-change-sensitive parameters (temperature, pH, gas phase composition and microbial activity) on inorganic pollutants mobility in the soil- and groundwater zone of the Dommel (Nl), Flémalle (B) and Gallego (E) sites (Joubert *et al.*, 2007). The distribution coefficient (Kd values) and transfer functions were calculated for As, Pb, Zn, Cd and Hg in the saturated and vadose zone from experimental results. Moreover, impacts of oxygen, nitrate and sulphate addition on biogeochemical functioning and metal release or precipitation/sorption reactions of saturated soils have been studied in laboratory experiments. These electron acceptors were tested because they can be introduced into the subsurface during events involving flooding and/or a fluctuating groundwater table that may result from climate changes (Vanbroekhoven *et al.*, 2007).

Further studies of bacterial biodiversity were performed with statistical analyses of soil incubations to show relationships between the bacterial community structure and arsenic turnover. Molecular approaches have been developed using functional genes to characterize bacterial biodiversity and activity involved in oxidation and reduction of arsenic at different sites in the Meuse river basin. This underlined the usefulness of the aoxB gene as a powerful molecular marker of aerobic As(III)-oxidizing bacteria and provided insight into their evolutionary relationships (Quéméneur *et al.*, 2007 and 2008).

The influence of environmental parameters (mainly acidic rainwater) on biogeochemical processes affecting the mobility of metals in the vadose zone was also studied in dynamic and long-term conditions using soil column experiments.

In the root zone, experiments in lysimeters under vegetation cover showed that heavy metal contamination induced more distributed plant water uptake along the soil profile compared to non-contaminated soil, as well as a deeper root penetration for contaminated soils. Further lysimeter experiments on the transport of lead in vegetated soil indicated preferential transport of lead and enhanced leaching via particle facilitated transport in root macro pores in environmental conditions in which lead is supposedly immobile (Roulier *et al.*, 2008).

Bioprocesses; organic contaminants

Microbiological work revealed functional activities and composition of microbial communities in the Brévilles, Danube, Ebro and Elbe. Molecular markers based on specific DNA and RNA sequences were applied that respond to system perturbations such as changes in redox conditions, water saturation and pollution. In Ebro river sediment the bacterial communities were shown to change with the seasons and are different on each sample location. The composition of the bacterial populations in the upper part (0–10 cm) of the sediment changed more over the seasons than that in the deeper (10–20 cm) sediment.

DNA chips with probes targeting genes involved in a variety of biogeochemically important processes (e.g., N-mineralisation, C-mineralisation, P-uptake, pollutant biodegradation and metal resistance) were used to detect different microbial functions in Ebro river sediment. The DNA chips revealed a high diversity of these microbial functions, but a clear relation to specific environmental conditions (temperature, pH, pollution, location) could not be established. Profiles of contaminant degradation genes, and more specifically those involved in anaerobic dehalogenation, changed significantly between sampling locations that differed in contaminant load, i.e., high contamination with hexachlorobenzene vs. diffuse pesticide pollution. Multivariance statistical analyses was applied to link variation in contaminant degradation genes to variation in geochemical conditions (Taş, 2009).

The biodegradation/biotranformation of a variety of organic compounds was tested, see Table 3.1. The degradation potential depends on the sediment and the conditions used, and this simplified table shows if degradation/transformation of the compound was observed without further details on the conditions applied.

As well as the in-situ field conditions assessment, laboratory setups were developed and tested to investigate stimulated conditions that can be applied in the field. The laboratory model systems included microcosms, mesocosms, chemostats and sediment columns to simulate various perturbations under environmental conditions such as changing temperature, redox, water saturation and influx of pollutants as typical environmental changes. The quantitative presence of functional microbial groups, like denitrifying, sulphate-reducing, methanogenic and dechlorinating micro-organisms, was tested in cultures through MPN-dilution experiments. Results were qualitatively comparable with results obtained by culture independent molecular methods, however molecular techniques had a higher resolution.

Table 3.1 Biodegradation/biotransformation potential of the tested organic compounds (+ = degradation potential present, – = no degradation potential observed, nt = not tested).

	Conditions			Conditions	
	Aerobic	Anaerobic		Aerobic	Anaerobic
Halogenated compounds			Pesticides		
1,2-Dichloroethane	+	+	Acetochlor	+	Nt
Hexachlorobenzene	nt	+	Atrazine	+	–
Trichlorobenzenes	nt	+	Isoproturon	+	Nt
Dichlorobenzenes	nt	+	Mecoprop (MCPP)	+	Nt
Hexabromocyclododecane	nt	+	beta-HCH		+
Perchloroethylene (PCE)	nt	+	3,4-Dichloroaniline	+	?
(Poly)cyclic aromatics			*Emerging compounds*		
Benzene	+	+	Nonylphenol	+	±
Naphthalene	+	+	Estrogenic compounds	+	nt
Acenaphthene	+	+	MtBE	+	–
Acenapthylene	+	+			

The dechlorinating bacteria *Dehalococcoides*, *Dehalobacter* and *Desulfitobacterium* were detected and quantified in the degradation studies. *Dehalococcoides* species were particularly abundant (in several cases 0.8% and higher) in the presence of halo-organic pollutants (Taş *et al.*, 2009; van der Zaan *et al.*, 2009) suggesting that their presence can be used as an indicator of contamination, and as an indicator of contaminant degradation. Moreover, it could be shown that dechlorinating bacteria were active over a range of environmental conditions, including different temperatures and salinities, important variables in estuarine catchments with strong seasonal dynamics.

Investigations into the mineralization potential of herbicides (isoproturon, acetochlor and MCPP) in the limestone catchment Brévilles revealed a substantial mineralization in the topsoil (19–44%) (Janniche *et al.*, 2009). The three pesticides were also mineralized in the aquifer material, 10 to 58 mbs, but to a lesser degree and varied in degradation rate and potential at different depth. This clearly demonstrated spatial variation within these geological settings, but the governing parameter for initiating mineralization of acetochlor and isoproturon is still unknown (Janniche *et al.*, 2007).

As biodegradation capacity may be different in the field than in laboratory studies and degradation rates can vary strongly in space and time, novel methods to quantify biodegradation based on stable isotope analysis were developed and tested. The methods include an assay where ^{13}C-labelled compounds are added (Morasch *et al.*, 2007) as well as a method based on changes of the stable isotope ratios of pollutants that can be integrated into a numerical modelling framework (Batlle-Aguilar *et al.*, 2009). Application of the methods to two sites contaminated with benzene (Batlle-Aguilar *et al.*, 2009) and chlorinated ethenes (Abe *et al.*, 2009) provided new insight into factors that control contaminant biodegradation at the groundwater-surface water interface.

The availability, biodegradability and toxicity of nonylphenol in contaminated river sediment of a tributary of the Ebro River in Spain was determined and these were correlated to the estrogenic toxicity of the sediment. Our results show that nonylphenol can be used as an indicator compound to monitor estrogenic toxicity of this Ebro River sediment (de Weert *et al.*, 2008).

A continuous reactor system was used to mimic a river with nonylphenol-polluted sediment (Figure 3.6). Both a sediment bed and resuspended sediment conditions were chosen to simulate a gently flowing river and a flooding, respectively. Our results show that the microorganisms present are able to degrade the desorbing nonylphenol from a sediment bed, which reduces the concentration in the bulk water. However, resuspension led to an immediate increase of nonylphenol released from the sediment. This released amount was higher than the amount that can be degraded by the microorganisms with an optimal biodegradation rate. Therefore, resuspension will lead to an increased nonylphenol concentration in the bulk water, and the environment becomes more polluted with the compound (de Weert *et al.*, 2010).

IMPLICATIONS AND APPLICATION OF RESULTS

Soils and sediments are primary sinks for a wide range of organic pollutants, and bioavailability of the pollutants can be reduced through adding of respective adsorbents. Sorption studies should focus on low (environmental relevant) concentrations, since derived parameters from batch experiments with high concentrations will possibly overestimate pore water concentrations by approximately one order of magnitude.

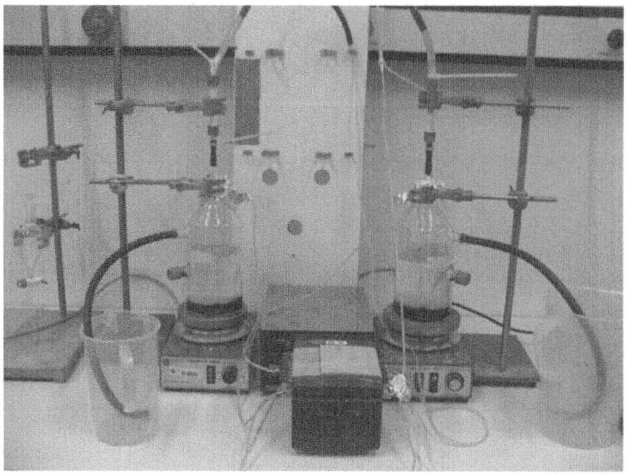

Figure 3.6 A continuous reactor system was used to mimic a river with nonylphenol polluted sediment.

Furthermore the river basins exhibit a remarkable biodegradation/biotransformation potential under different redox conditions tested. This potential may be used to enhance the removal or immobilization of both inorganic and organic pollutants provided suitable conditions are applied through appropriate river basin management.

CONTRIBUTIONS AND POLICY RECOMMENDATIONS FOR RIVER BASIN MANAGEMENT

Managing pollutants at the basin scale means regulation of pollutant sources. Hence, identification of pollutant sources is still mandatory (see also chapters MONITOR and BASIN). To reduce the mobility of pesticides their application should be limited to distinct weather conditions (i.e. during a dry period but after a rain event when soil water content is high but groundwater recharge is low). More information on the degradation under specific conditions is important, but even the slightest degradation or sorption has an impact on the presence of the pollutants in the environment and should therefore be taken into account.

River basin management needs to take into account the distribution of floodplain contaminants when designating areas for re-naturation and flood control storage. Summer flooding may lead to temporary reducing conditions during which mobilization of certain contaminants (e.g., arsenic) but also transformation of certain organic compounds may be drastically enhanced. As basin scale models rarely include redox dynamics and colloid transport, our findings suggest that actual contaminant mobility might be well higher compared to current model predictions.

Furthermore, a flooding can also result in more polluted environments, e.g. if the desorption of organic contaminants is higher than the biological degradation. This creates an environment with a higher temporary risk than the one before.

The studies on biodegradation of contaminants at the groundwater-surface water interface revealed that rates of contaminant attenuation can strongly vary between the aquifer and the zone close to the river due to changes in geochemical conditions. Therefore, when evaluating the risk of impacts of groundwater contamination on surface water and groundwater-dependant ecosystems, attenuation processes in these zones should be evaluated separately. Furthermore, since biodegradation rates vary strongly between sites, whenever possible site-specific degradation rates and site-specific measures to enhance degradation should be determined.

Locally/nationally

The field studies were carried out in close contact with regulators and consultants, which showed a large interest in incorporating some of the developed tools to assess natural attenuation into the risk assessment procedure.

Our findings indicate that colloid-facilitated transport affects the speciation, distribution, and bioavailability of contaminants within the floodplain ecosystem and their transport behaviour along the river basin.

Carbonaceous materials such as activated carbon act as strong adsorbents for organic pollutants as they have the potential to immobilize them. Hence, such materials can be applied on contaminated sites for sequestration of the respective pollutants. Such amendment of activated carbon is currently under discussion as a cost-effective remediation strategy compared to conventional approaches. However, one needs to take into consideration that addition of a strong adsorbent will reduce the bioavailability of pollutants thereby limiting the potential for natural or enhanced attenuation of these compounds.

The results from pesticide mineralization and sorption investigations lead to the conclusion that a limestone catchment like Brévilles is very vulnerable towards pesticides and has a substantial risk for groundwater contamination with the applied pesticides. These types of geological settings are thus unable to protect river basins from contaminants applied to agricultural area in the catchment. It would therefore be recommendable to consider the underlying geology of the agricultural fields when allowing pesticide use in agriculture.

In a European context

Experiments on soil flooding performed at different temperature demonstrated the effect of the soil reduction kinetics on contaminant mobilization. These results suggest that the impact of climate change on contaminant mobilization from riparian floodplains occurs via frequency and duration of flooding, timing and temperature of flooding (e.g., summer vs. winter floods), and possibly secondary effects including the quality of microbial degradable carbon and the amount of sulphate available for soil reduction and colloid formation.

The poly-parameter linear free energy relationships provide a robust tool for a rapid first prediction of sorption parameters based on intrinsic compound properties without any measurement. Such *a priori* estimations are extremely useful for the determination of the "favoured" compartment of compounds that were so far not studied, e.g. emerging pollutants.

Transfer functions (or Kd functions) link the Kd of each inorganic pollutant to biogeochemical characteristics of soils using mathematical equations. By coupling the Kd analysis with microbial and geochemical parameters, the potential of the soil to release inorganic pollutants can be evaluated.

The evidence that even strongly adsorbing compounds such as Pb may leach in the root zone (preferential flow and particle facilitated transport) highlights the need to consider a "dynamic" adsorption coefficient for the root zone. The modelling approach described here can be used to estimate such a coefficient. Results of heavy metal leaching experiments in lysimeters with different vegetation cover may assist in the development of managing strategies for metal contaminated catchments using plants.

Molecular identification of microbial communities can be used to identify the microbial characteristics of soils, sediments and surface water and is a monitoring tool to assist system trend prediction and system mitigations. Measurements of microbial numbers and activities can be used to assess intrinsic and enhanced natural attenuation as they show the amount or activity of present bacteria, which can be compared to reference values in time or in space.

Biodegradation rates of pollutants were determined in the laboratory (laboratory tests) and directly in the field (stable isotope analyses). These degradation rates can be used to predict whether a compound will be present in the environment over a long period of time, or whether the system can deal with it naturally and transform it to harmless compounds. Even the lowest degradation rate can have an important impact on the ecosystem.

Bioavailability analysis and toxicity tests of emerging compounds such as nonylphenol and other estrogens show a coupling between biodegradation, bioavailability and toxicity. Furthermore, concentration measurements of nonylphenol can be used as a marker to predict the estrogenic toxicity of sediment, thus complex analyses of all estrogenic compounds or toxicity tests for all sampling points or time intervals is no longer necessary for an adequate assessment.

The findings on the fate of organic contaminants at the groundwater-surface water interface are highly relevant to better understand contaminant fluxes to groundwater-dependant ecosystems, which is an important issue of the GWD and an area in which little knowledge was available so far.

OPEN & NEW QUESTIONS FROM BIOGEOCHEM

What is the effect of sulphide colloid formation on mercury dynamics? The effects of sulphide colloid formation on mercury dynamics (which is an important contaminant within the Elbe basin floodplains) was not evaluated so far. However, based on mercury chemistry, similar colloidal mobilization as was observed for copper can be expected. Further research is currently underway to clarify this question.

How mobile are metal sulphide colloids in the floodplain soils? While copper sulphide colloids appear to be mobile in mesocosm experiments, long-range colloid transport needs further attention to assess contaminant migration from floodplains draining to groundwater or exfiltrating into surface water.

Can added carbonaceous adsorbents stabilize the organic pollutants in the environment? Based on our results that organic contaminants are associated with strong adsorbents like carbonaceous materials, the abundance of anthropogenic carbonaceous materials in river basins can stimulate the immobilization of organic contaminants. The half-life times of the contaminants will be strongly affected, and will this be a long-lasting effect? For instance, co-sorption of dissolved organic matter to carbonaceous adsorbents can reduce their sorption capacities. Furthermore, the stability of carbonaceous adsorbents in different environments needs to be studied (i.e., time-scale of weathering of such materials).

Is it possible to extrapolate the results on transfer functions and bacterial bio-indicators to other sites? This method was demonstrated and validated with sediment from three different sites, and should be validated with other sediments as well to make it a general applicable tool.

How can the long-term risk of heavy metal contaminated soils be assessed with transport models, the plant cycle in relation to the production of organic compounds and preferential flow paths? During our study we noticed shortcomings of whole system simulation models in reproducing lead leaching in the root zone of a vegetated soil. This highlights the importance of the plant in the long-term mobility of strongly adsorbing compounds (Roulier *et al.*, 2008).

Can we use molecular markers for other compounds in the field? This technology was demonstrated for dechlorination processes, based on the knowledge on degradation pathways, responsible bacteria and involved enzymes. This technology can also be developed for other degradation processes.

Can we quantify degradation rates in the field for other compounds? Quantifying reliable and realistic field degradation rates becomes more and more important for modelling purposes. The use of stable isotopes to calculate them was demonstrated for other aromatic hydrocarbons and chlorinated ethenes, but can also be developed for other contaminants, such as pesticides?

Can nonylphenol be used universally as marker for estrogenic toxicity? This technology was demonstrated and validated with Ebro sediments, and should be validated with other sediments as well to make it a general applicable tool.

ACKNOWLEDGEMENT

We thank all members of the BioGeoChemistry workpackages of the Aquaterra project for their contribution to this Chapter: Hans-Jorgen Albrechtsen, Nicole Baran, Johannes Barth, Dominique Breeze, Catherine Crouzet, Jasperien de Weert, Ludo Diels, Kieron Doick, Igor Dubus, Satoshi Endo, Anne Gautier, Jan Gerritse, Peter Grathwohl, Ilka Herklotz, Anke Hofacker, Gry Sander Janniche, Michel Jauzein, Kevin C. Jones, Antoine Joubert, Catherine Joulian, Bertrand Ligouis, Lise Lucas, Claudia Moeckel, Sandra Meijer, Barbara Morasch, Christophe Mouvet, Britt Pagels, Marianne Quéméneur, Stéphanie Roulier, Alain Saada, Gosse Schraa, Kirk Semple, Piet Seuntjens, Hauke Smidt, Dietmar Steidle, Nicolas Surdyk, Neslihan Tas, Kai Totsche, Stephane Touzelet, Karolien Vanbroekhoven, Bas van der Zaan, Miriam van Eekert, Sandra Van Roy, Andreas Voegelin, and Frank-Andreas Weber.

REFERENCES

Abe, Y., Aravena, R., Zopfi, J., Parker, B.L. and Hunkeler, D. (2009) Evaluating the fate of chlorinated ethenes in streambed sediments by combining stable isotope, geochemical and microbial methods. *J. Contam. Hydrol.* **107**(1), 10–21.

Abelmann, K., Kleineidam, S., Knicker, H., Grathwohl, P. and Koegel-Knabner, I. (2005) Sorption of HOC in soils with carbonaceous contamination: Influence of organic-matter composition. *Journal of Plant Nutrition and Soil Science* **168**, 293–306.

Batlle-Aguilar, J., Brouyère, S., Dassargues, A., Morasch, B., Hunkeler, D., Höhener, P., Diels, L., Vanbroekhoven, K., Seuntjens, P. and Halen, H. (2009) Benzene dispersion and natural attenuation in an alluvial aquifer with strong interactions with surface water. *J. Hydrol.* **369**, 305–317.

de Weert, J., Streminska, M., Hua, D., Grotenhuis, T., Langenhoff, A. and Rijnaarts, H. (2010) Nonylphenol mass transfer from field aged sediments and subsequent biodegradation in reactors mimicking different river conditions. *Journal of Soils and Sediment* **10**, 77–88.

de Weert, J., De la Cal, A., van den Berg, H., Murk, A., Langenhoff, A., Rijnaarts, H. and Grotenhuis, H. (2008) Bioavailability and biodegradation of nonylphenol in sediment determined with chemical and bioanalysis Enviromental Toxicology and Chemistry. **27**(4): 778–785.

Endo S., Grathwohl P., Haderlein S.B. and Schmidt T.C. (2008) Compound-specific factors influencing sorption nonlinearity in natural organic matter. *Environmental Science and Technology* **42**, 5897–5903.

Endo S., Grathwohl P., Haderlein S.B. and Schmidt T.C. (2009a) Characterization of sorbent properties of soil organic matter and carbonaceous geosorbents using n-alkanes and cycloalkanes as molecular probes. *Environmental Science and Technology* **43**, 393–400.

Endo S., Grathwohl P., Haderlein S.B. and Schmidt T.C. (2009b) LFERs for soil organic carbon – water distribution coefficients (K_{OC}) at environmentally relevant sorbate concentrations. *Environmental Science and Technology* **43**, 3094–3100.

Janniche, G. S., Baran, N., Gutierrez, A., Mouvet, C. and Albrechtsen, H.-J. (2009) Fate of herbicides in Brévilles catchment from topsoil through aquifer to spring. Conference Proceedings Aquaterra Final conference – Processes – data – models – future scenarios. Scientific fundamentals for river basin management, March 2009, Tübingen, Germany, p. 23.

Janniche, G. S., Lindberg, I.E., Mouvet, C. and Albrechtsen, H.-J. (2007) Spatial small-scale variation in degradation of pesticides (acetochlor, isoproturon MCPP) in limestone and sand from Brévilles, France. Conference Proceedings International workshop on "Biogeochemical processes in groundwater systems", GSF Campus, September 2007, Munich, Germany.

Joubert, A.V.P., Lucas, L., Garrido, F., Joulian, C. and Jauzein, M. (2007) Effect of temperature, gas phase composition, pH and microbial activity on As, Zn, Pb and Cd mobility in selected soils in the Ebro and Meuse Basins in the context of global change. *Environmental Pollution* **148**, 749–758.

Morasch, B., Hohener, P. and Hunkeler, D. (2007) Evidence for insitu degradation of mono-and polyaromatic hydrocarbons in alluvial sediments based on microcosm experiments with 13C-labeled contaminants. *Environmental Pollution* **148**, 739–748.

Navarro, A., Endo, S., Gocht, T., Barth, J., Lacorte, S., Barceló, D. and Grathwohl, P. (2009) Sorption of alkylphenols on Ebro River sediments: Comparing isotherms with field observations in river water and sediments. *Environ. Poll.* **157**, 698–703.

Quemeneur, M., Garrido, F., Jauzein, M. and Joulian, C. (2007) AoxB gene as molecular marker of arsenite-oxidizing bacteria., in BAGECO9 – 9th Symposium on Bacterial Genetics and Ecology – Wernigenode – Germany – 23–27/06/2007.

Quemeneur, M., Joubert, A., Garrido, F., Jauzein, M. and Joulian, C. (2008) Bacterial community dynamic and potential arsenic biotransformation in industrial contaminated soil leachates., in ConSoil 2008 – Milan – Italie – 03–06/06/2008.

Roulier, S., Robinson, B. Kuster, E. and Schulin, R. (2008) Analysing the preferential transport of lead in a vegetated roadside soil using lysimeter experiments and a dual-porosity model. *European Journal of Soil Science* **59**, 61–70.

Taş, N. (2009) Dehalococcoides spp. in river sediments: Insights in functional diversity and dechlorination activity, PhD thesis Wageningen University, Wageningen, p. 135.

Taş, N., van Eekert, M.H.A., Schraa, G., Zhou, J., de Vos, W.M. and Smidt, H. (2009) Tracking functional guilds: Dehalococcoides spp. in European river basins contaminated with hexachlorobenzene. *Applied and Environmental Microbiology* **75**, 4696–4704.

Totsche, K.U., Jann, S. and Kögel-Knabner, I. (2007) Single event-driven export of polycyclic aromatic hydrocarbons and suspended matter from coal tar-contaminated soil, *Vadose Zone J.* **6**(2), 233–243.

van der Zaan, B.M., de Weert, J., de Vos, W.M., Rijnaarts, H., Smidt, H. and Gerritse, J. (2009) Degradation of 1,2-dichloroethane by microbial communities from river sediment at various redox conditions. *Water Research* **43**(13), 3207–3216.

Vanbroekhoven, K., Van Roy, S., Gielen, C., Maesen, M., Ryngaert, A., Diels, L. and Seuntjens, P. (2007) Microbial processes as key drivers for metal (im)mobilization along a redox gradient in the saturated zone. *Environmental Pollution* **148**, 759–769.

Voegelin, A., Weber, F.-A. and Kretzschmar, R. (2007) Distribution and speciation of arsenic around roots in a contaminated riparian floodplain soil: Micro-XRF element mapping and EXAFS spectroscopy. Geochimica et Cosmochimica Acta, 5804–5820.

Weber, F.-A., Voegelin, A., Kaegi, R. and Kretzschmar, R. (2009) Contaminant mobilization by metallic copper and metal sulphide colloids in flooded soil. *Nature Geoscience* **2**, 267–271.

Chapter 4

FLUX – Mass transfer in river-sediment-soil-groundwater systems

Ph. Négrel, E. Petelet-Giraud, G. Klaver, A. Marsman, K. C. Jones, P. Middeldorp, N. Baran, A. Brenot, T. Gocht, P. Grathwohl, J. P. Hermand, E. Kalbus, S. Meijer, R. Millot, C. Moeckel, F. Reinstorf, S. Roy, M. Schirmer, C. Schmidt, D. Steidle, B. van Os and M. Verbanck

KEY OBJECTIVES

The subproject FLUX aimed at the identification the impact of global and especially climate change on the functioning of 'river-sediment-soil-groundwater' systems. Fluxes of solids and solutes (inorganic and organic) in the water cycle depend on soil properties (filter, buffer, transformation capacity). The role of soils as the most active compartment in the water cycle has to be assessed in order to anticipate future trends deteriorating water quantity and quality. This was based on input/output mass balances of solids, solutes and pollutants in sub-catchments in close connection to the river basin investigations. Furthermore, inter-compartment fluxes (soil to groundwater/groundwater to surface water and vice versa) was addressed. The work planned requires the close collaboration of scientists from hydrology, inorganic geochemistry and environmental chemistry at selected field sites. The results were needed for the validation of integrated soil-water numerical models. FLUX consisted of three work packages, which targeted the following objectives.

In ***FLUX 1***, objectives were:

- To determine mass balances of pesticides at the scale of a small catchment (Brévilles, BASIN study case R1) where land use results in significant contamination of the groundwater.
- To determine the current burden of atrazine and desethylatrazine in soils and sub-soils of the Brévilles catchment, in a way that informs the quantification of the residual mass of these compounds that can lead to the observed continued contamination of the groundwater 5 years after the last application of atrazine.
- To determine the net air-soil exchange direction and fluxes of a range of persistent organic pollutants at the local and catchment scale.
- To determine the current burden in soils at the appropriate study catchment scale and the distribution within soils, in order to gain understanding of processes controlling air-surface exchange and soil sub-surface supply.
- To utilise the knowledge obtained from the above, and from other components of the project, to assess likely future trends in pollutant loadings to surface soils, surface water and air.

In *FLUX 2*, objectives were:

- Obtain a generic intercompartment flux model to assess fluxes from groundwater to surface water and vice versa on a river subcatchment and catchment scale.
- Determine trends in the solutes discharge into the river system and discharge from the river system into the groundwater as a result of perturbations, i.e., the combined effect of anthropogenic contamination and climate change.
- To get insight in the role of the interface between groundwater and surface water in the decrease of contaminant fluxes in catchment bodies (flow path soil – groundwater – surface water and vice versa).
- Establish the mass balance and fate of solutes at the selected river basin scale (ca. 125 km^2), especially at the infiltration/seepage area from groundwater to surface water and vice versa. These intercompartment fluxes will be studied and quantified for different representative river catchment systems.

In *FLUX 3*, objectives were:

- Determination of an output-input budget with the determination of element origins in the dissolved load through the different geochemical signatures of the several end-members (e.g., natural: rock weathering; anthropogenic: agricultural and urban).
- For solid matter, with regard to temporal fluctuations, determination of the origin of the suspended matter through the different geochemical signatures which will be related to the different parts of the Dommel and Meuse basins.
- Coupling present and paleo-fluxes in order to make future flux predictions. In order to assess the physical, chemical and biological properties of the sediment caused by changing climatic conditions, land use practices and other perturbations, the paleo-state and history of those changes must be known. Developing a robust methodology for quantifying particle fluxes at the basin scale. Assessing the sequence of sediments and their changes in riverbeds by novel acoustic methods.
- Restoring the landscape: will the future be as good as the past? Realistic estimates for restoring ecological areas and reconstruction of the pre-anthropogenic ecological landscape on the basis of the sediment record. Which scenario's are realistic when restoring such landscapes, based on the future flux predictions.

KEY ACHIEVEMENTS

Distribution and fluxes of persistent organic pollutants in AquaTerra catchments

In FLUX 1, the fluxes and long-term fate of persistent organic pollutants (POPs) were estimated in various catchments. A database for evaluation and interpretation of net exchanges/fluxes has been made available containing pollutant concentrations in air (from passive air samplers), atmospheric deposition, water and soil samples in the catchments of Danube, Elbe, Ebro and Brévilles. Based on these data, diffusive soil-air fluxes were calculated for PCBs at selected sites in the Ebro, Danube and Elbe basins. Most sites showed a net deposition of PCBs to the soil, with the exception of the most volatile compounds, i.e., PCB 28 (Figure 4.1).

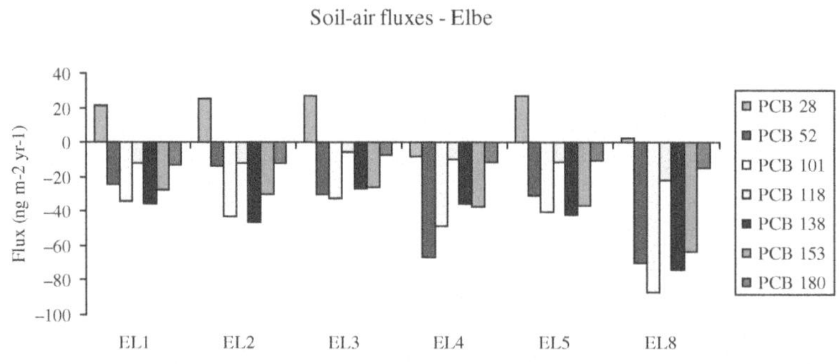

Figure 4.1 Estimated net soil-air fluxes in the Elbe basin (positive fluxes mean net volatilisation).

A regionally segmented multimedia fate and transport model (Evn-BETR model, Prevedouros *et al.*, 2004) was used to estimate the relative importance of volatilization, degradation and leaching with respect to losses of PCB 28, 153 and 180 from soils. The model predicted that degradation is the most important loss process for all PCBs, while runoff to freshwater is also important for the higher-molecular weight PCBs (PCB 153 and 180 up to 99% by chemical bound to particles) Movement to deeper soil layers, root uptake by vegetation and volatilization were all predicted to be unimportant loss processes. However, the model does not consider DOC- or colloid-facilitated transport of chemicals.

However, measurements of PCBs in different soil layers showed possible evidence for leaching of tri- and tetra-chlorinated PCBs.

By measuring PAHs in soil and stream water, estimates of soil storages and water fluxes can be made and compared with atmospheric deposition. The results indicate that more than 90% of the incoming PAHs remain in the catchments and accumulate in the top-soils, with re-volatilization of PAHs degradation in the soils being very limited. The reason for the persistence of atmospheric PAHs in rural soils is very likely their strong adsorption to combustion-derived carbonaceous material.

Atmospheric deposition of PAHs was also compared to PAH soil burden at a site in the Danube Catchment. Although the average atmospheric deposition rate in the Danube Catchment is the highest among the five investigated European catchments (482 ng m^{-2} d^{-1}), the PAH concentration in the soil is comparably low (~200 µg kg^{-1}). Similar PAH soil values were found in the Brévilles and Ebro Catchments, but not in the Elbe Basin, where the concentrations are 10–100 times greater.

Fluxes of current use pesticide in the Brevilles catchment

The annual input to the soil of the most relevant and currently used pesticides has been established by farmers' interviews for the period 1994–2006 at the Brévilles catchment. Pesticide fluxes were calculated based on the monitoring of the flow and quality of the Brevilles Brévilles spring, which represents the major outlet of the system. The mass balances for atrazine and its major metabolite deethylatrazine, which are the only molecules systematically detected, reveal a loss at the major spring lower than 0.5% of the applied quantity of atrazine during the last decade. This implies that more than 99% of the applied pesticides either i) leave the basin via a different route (export via plants, volatilisation), ii) are stored in the soil, the unsaturated or saturated zone, iii) are degraded into metabolites not studied, or iv) are still in transfer and haven't reached the spring. On the basis of our present knowledge, the hypotheses ii) and iv) seem the most likely. The concentrations of atrazine (At), desethylatrazine (DEA), desisopropylatrazine (DIA) have been measured in representative soils of Brévilles watershed. DEA and DIA were below the limit of detection in all soil samples, whereas the At concentrations ranged from 5 ng/g to below detection limit. Based on these measurements and data available on the surface area of each plot, the soil thickness sampled, and the soil density, it was estimated that the total cumulated mass of At in the soils was 1 kg for the whole watershed. A fraction of this mass (determined 7 years after the last application of At) is possibly available for leaching and could generate a water infiltrating with At and/or DEA concentrations higher than the drinking water limit (0.1 µg/L). Thus, it is to be feared that a return of the ground water to a pristine condition could be very long probably on the order of decade even several decades.

In FLUX 2, fluxes of current use pesticide were determined in the Brevilles catchment with the help of isotope tools. The conceptual scheme of the Bitterfeld/Wolfen megasite, Mulde catchment, Elbe Basin, was constrained by isotopes tools while in the same area water and contaminant fluxes between aquifer, streambed and stream were quantified using integral pumping tests, streambed temperature mapping and streambed sediment sampling.

Geochemical conceptual scheme of the Bitterfeld/Wolfen megasite, Mulde catchment, Elbe Basin

The Bitterfeld/Wolfen region is a megasite with multiple contaminant sources from more than a century of industrial activity, which have a considerable impact on the environment. At present, the contaminated groundwater covers an area of about 25 km^2 and poses a threat for the surrounding aquifers and the Mulde River. Focus was put on the Schachtgraben, a man-made channel in the Mulde Floodplain that collects the effluents of the industrial area. In this hotspot area we managed to characterise the relationship between surface water (channels, rivers) and the groundwater in the shallow Quaternary aquifer. This is of major importance to decide on pollutant transfer pathways and how to interfere with them. Waters are Ca-SO_4 type with TDS reaching 3.8 g L^{-1} in the industrial area. Stable isotopes ($\delta^{18}O$, δ^2H) show that two of the rivers are recharged mainly by groundwater that can be divided into two groups. Strontium isotopes ($^{87}Sr/^{86}Sr$, Figure 4.2) designate different geochemical end-members and enable the identification of mixing between natural and anthropogenic surface and groundwater (Petelet-Giraud *et al.*, 2007).

Temperature as intercompartment water and contaminant fluxes characterization

In the same area water and contaminant fluxes between aquifer, streambed and stream were investigated using integral pumping tests, streambed temperature mapping and streambed sediment sampling (Figure 4.3). While the contamination in the groundwater was found to be diffuse the contaminant concentrations in the groundwater and in the stream water are considerably lower than in the pore water of the streambed sediments. The sediments constitute a contaminant source which may not be depleted in the near future due to very low groundwater discharge rates in some streambed sections. Thus, contaminant mass fluxes into the stream may continue for a long time period and pose a threat to the stream ecosystem.

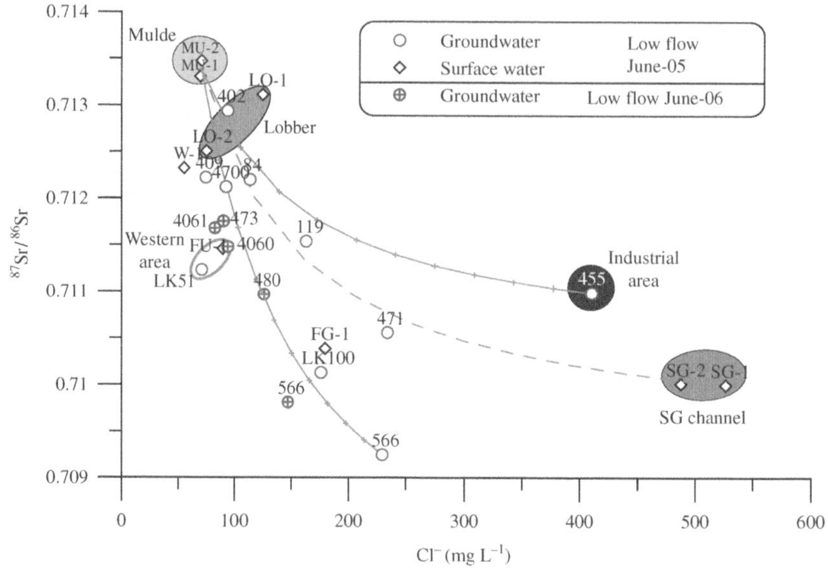

Figure 4.2 Plot of $^{87}Sr/^{86}Sr$ vs. Cl in surface and groundwater collected from the study site during low flow (with calculated mixing curves).

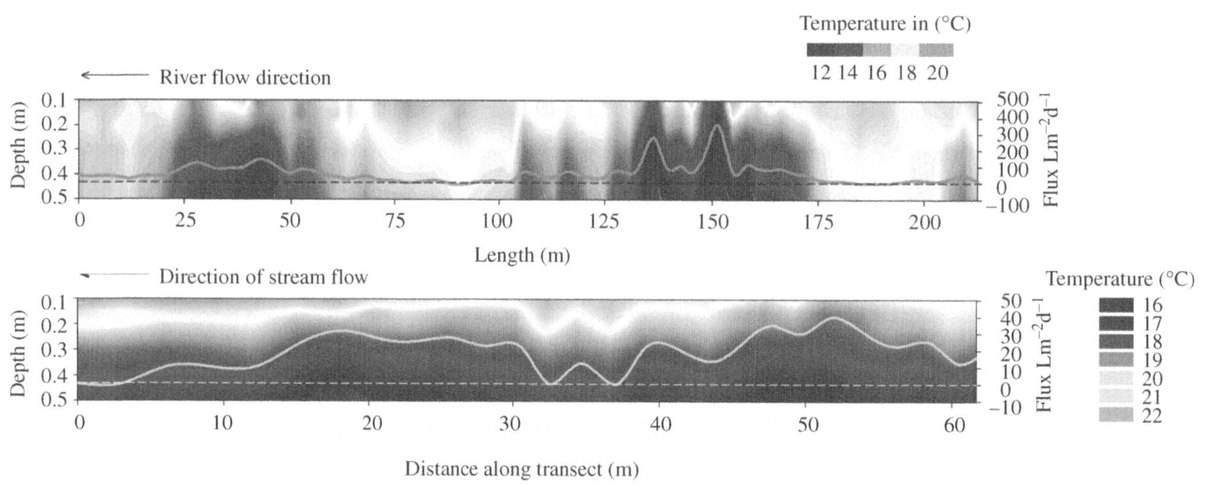

Figure 4.3 Temperature distribution in the streambed and calculated groundwater fluxes through the streambed in a) Section S and b) Section K.

Natural degradation processes at the interface between groundwater and surface water at laboratory scale

Particularly at the interface between ground- and surface water, natural degradation processes take place that eliminate mobile, aerobically degradable pollutants. For instance, infiltration of oxygen causes natural degradation processes to start. The scale of this process is a millimetre scale and therefore modelling results are calibrated with laboratory experiments on the same scale.

Detailed oxygen measurements using a micro-electrode system with micro-manipulator have shown that oxygen diffusion from surface water into the sediment in a simulated steady state system is on the order of millimetres (4–6 mm). Both laboratory experiments and model calculations show that an increasing seepage velocity causes a decrease of the oxygen penetration depth. The MCB flux reduction in a steady state situation at 20°C due to passage through the NA-interface was at least 70% and may be up to 98% as observed in the laboratory column simulations. Biodegradation of monochlorobenzene in the Solvay ditch in Amsterdam occurs also at a high rate in the surface water, which makes the distinction in the field between sediment and surface water difficult. This renders the relevance of the current policy of zero-emission from groundwater toward the surface water questionable. Flux reduction measurements of MCB in the field showed that the MCB removal was less in the winter season than in the summer season.

Geochemical conceptual scheme of the Brévilles catchment

Intensive monitoring of the Cuise Sands Aquifer in an area of 3 km^2, over nearly 5 years showed significant disparities for major element concentrations, Sr concentrations, isotopic compositions and ^3H values, depending on the groundwater sampling locations. Lower $^{87}Sr/^{86}Sr$ values associated with low NO_3 concentrations, low Mg/Sr molar ratios (<100) and low ^3H values are attributed to natural end-members. Groundwater with higher $^{87}Sr/^{86}Sr$ values and higher Mg/Sr molar ratios (>250) suggest a contribution of Sr from anthropogenic sources. Strontium isotopes coupled with Mg/Sr and Ca/Sr ratios of groundwater in this catchment successfully demonstrated the capacity to trace fertilizer contributions, which significantly modify the natural Sr isotopic signature of groundwater (Figure 4.4).

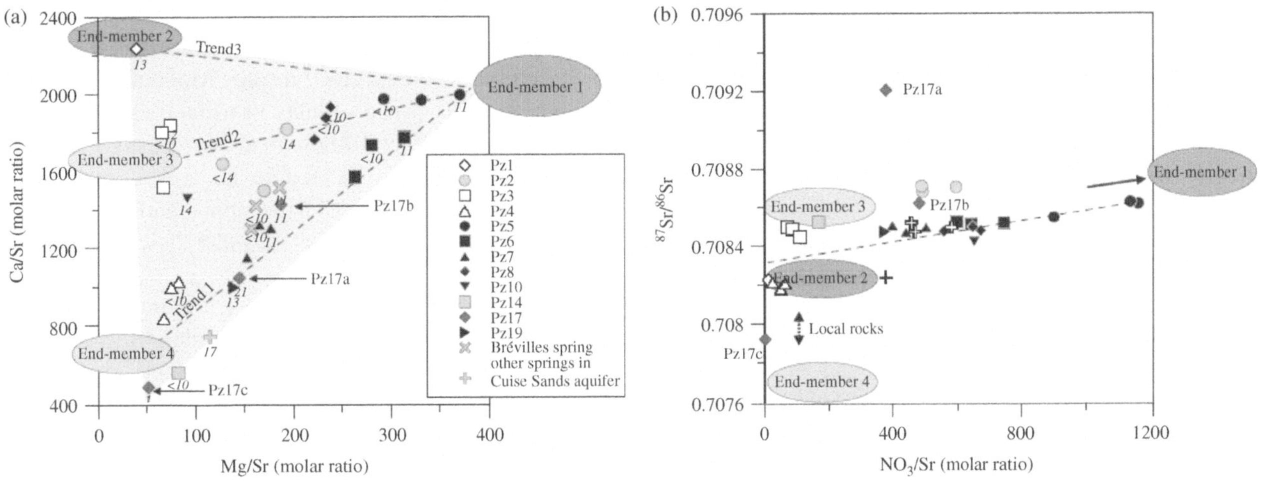

Figure 4.4 Geochemistry of groundwater collected in piezometers and springs in the Brévilles basin: (a) Ca/Sr vs. Mg/Sr; (b) $^{87}Sr/^{86}Sr$ vs. NO_3/Sr.

The specific geochemical signature of groundwater revealed by Sr isotopes, in addition to major and trace element concentrations (in particular Mg/Sr and Ca/Sr molar ratios), demonstrated the partial isolation of groundwater bodies in this catchment. Compartmentalisation of the aquifer is compatible with fault distribution as defined by geological and geophysical investigations. Thus, the Sr isotopic tool coupled with Mg/Sr and Ca/Sr ratios has been applied with success to trace water pathways and recharge and to provide scenarios for the hydrological functioning of the system (Brenot *et al.*, 2008).

In FLUX 3, output-input budget with the determination of element origins in the dissolved load have been investigated in the Ebro and Dommel catchments. For solid matter, with regard to temporal fluctuations, determination of the origin of the suspended matter through the different geochemical signatures has been investigated in the different parts of the Dommel and Meuse basins.

Influence of dams on the composition of the suspended and riverbank sediments in the Danube

We found that compared with undammed Danube sections, Iron Gate reservoir sediment and suspended matter showed higher clay contents and different K/Ga and Metal/Ga ratios. These findings emphasize the importance of reservoir-river sediment-fractionation (Klaver *et al.*, 2007).

Long-term fluxes of dissolved and suspended matter in the Ebro River Basin

The spatial and temporal distribution of physical and chemical parameters, suspended particulate matter (SPM) and major elements were investigated in the Ebro Basin upstream in Mendavia and downstream in Tortosa between 1981 and 2004 using the historical databank. The mean interannual dissolved fluxes were estimated to be 1.6 10^6 $t.y^{-1}$ at Mendavia and 5.9 10^6 $t.y^{-1}$ at Tortosa respectively, leading to evaluated specific export rates close to 93 $t.km^{-2}.y^{-1}$ at Mendavia and 69 $t.km^{-2}.y^{-1}$ at Tortosa respectively. The SPM flow related to the specific mechanical erosion rate over the whole basin was around 1.0 10^5 $t.y^{-1}$, providing an estimate of the specific export rates of around 5.1 $t.km^{-2}.y^{-1}$ at Mendavia and 1.2 $t.km^{-2}.y^{-1}$ at Tortosa, thus highlighting the control of solid transport by the numerous dams within the Ebro River Basin. Calculations were also carried out for the respective contribution of carbonate and evaporite (gypsum) with respect to the total dissolved solids (TDS). It appears that the TDS, over the whole catchment scale, is mostly controlled by SO_4, Na and Ca concentrations, reflecting the preferential contribution of both evaporite and carbonate dissolution during chemical weathering within the Ebro River Basin (Négrel *et al.*, 2007).

Decomposition of dissolved elements fluxes on the Ebro River Basin (Spain) supported by sulfate isotopes

The *Ebro River* is remarkable for its high mineralisation dominated by SO_4, Ca and Cl concentrations. The different tributaries of the Ebro present contrasting geochemical signature, reflecting disparities in lithologies and land use in the basin. The contribution of the different lithological sources (evaporites, carbonate and silicate) has been identified on the catchment, using the isotopic composition and the concentration in dissolved sulfates. These were related to space (lithology and land use) and time (mixing process and land use). Tributaries draining the Pyrenees Mountains upstream, integrate, in addition to evaporite sulfate, sulfates from sulfide oxidation. Dissolved sulfates in tributaries draining the southern part of the Ebro basin are only controlled by evaporite dissolution. To these lithological signals, anthropogenic input of S-fertilizers is overloaded depending on time and space. The Gallego located in an area dominated by agriculture, is strongly marked by S-fertilizer contribution. In addition we found that in the upper part of the basin, tributaries inputs are marked by significant groundwater contributions. Downstream, loss of water potentially due to evaporation process, canal seepage and irrigation has been evidenced.

The composition of the Ebro River at the outlet of the basin (Amposta) has been monitored on a monthly basis over one year. The observed seasonal variations of sources in dissolved sulfates (evaporite, sulfide oxidation, S-fertilizer) have been interpreted as various mixing of water supplies by different compartment (tributaries and groundwater contributions) on the river catchment. The seasonality of the anthropogenic inputs could either translate increasing contribution of surface run-off on the Ebro basin and/or increasing contribution of groundwater marked by S-fertilizer signature (Brenot *et al.*, submitted, Figure 4.5).

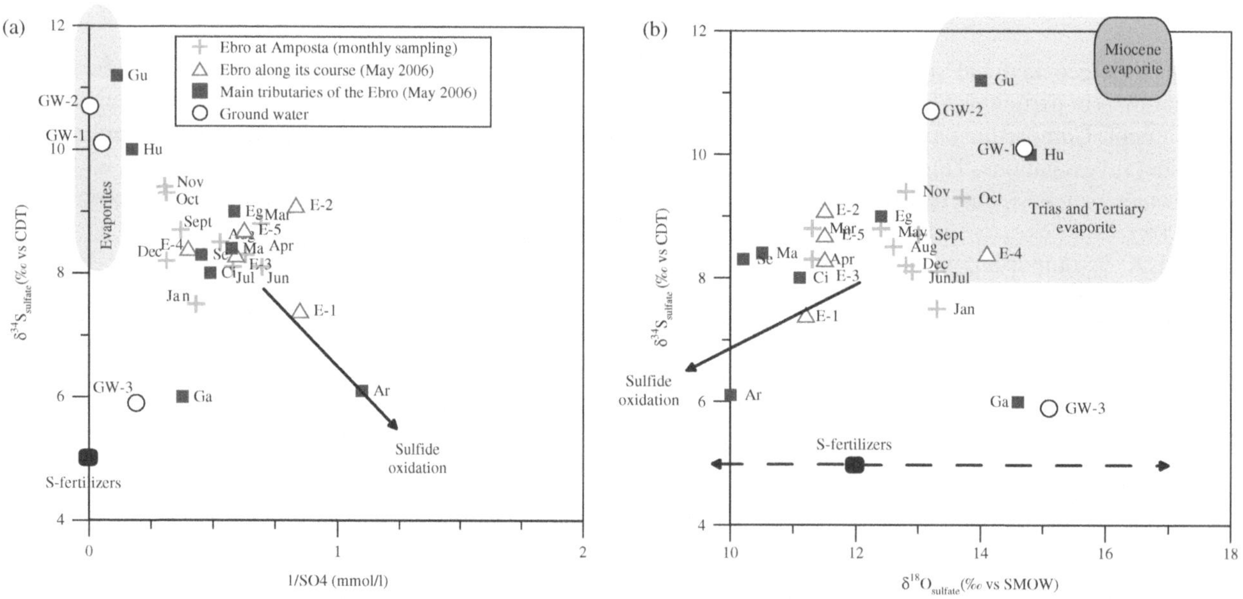

Figure 4.5 $\delta^{34}S_{sulfate}$ vs. 1/SO4 (a) and vs. $\delta^{18}O_{sulfate}$ (b) for river and groundwater sampled on the Ebro basin.

Natural versus anthropogenic sources in the Dommel surface and groundwater dissolved load (Meuse basin): isotopic constraints (B, Sr) and Gd anomaly

The *Dommel River*, a tributary of the Meuse River, drains an intensive agricultural and heavily populated area. In the border region of Belgium and the Netherlands, three zinc-ore smelters are located within 10 km. Exhaust fumes from these smelters have emitted oxides of heavy metals during about one century, resulting in cadmium and zinc contaminated soils in this region. Although since the seventies the atmospheric emissions have been greatly reducedagricultural activities changed from dairy farming to intensive livestock farming. The manure was spread on a relative small area of land and leads to a large load of nitrate, phosphate and also metals.

We could show that human activities in the Dommel catchment induce a large amount of dissolved elements released to surface and ground waters. The aim of our study was to discriminate the natural versus anthropogenic sources of the dissolved load and also to outline pollution sources including agriculture, industrial activities, wastewater treatment plant effluents). For this, a monthly monitoring of strategic locations together with an instantaneous geochemical picture of the whole catchment was realised. The use of combined geochemical tracers, i.e., strontium isotopes, boron isotopes and the gadolinium anomaly, revealed the main sources of contaminants in the basin as effluents of water treatment plants, agriculture and smelter influence (Petelet-Giraud *et al.*, 2009).

Heavy metal sources in the Dommel catchment (Meuse basin): Chemical and isotopic constraints

The soils in the Kempen region are anthropogenically enriched in Cd and Zn by the historical atmospheric pollution of four zinc smelters. During raining periods, the stream water is predominantly fed by shallow groundwater. This recent groundwater leaches highly polluted soils and leads to elevated Cd and Zn concentrations also in surface water. In contrast, during dry periods, surface water originates from unpolluted deep groundwater. The Dommel receives Cd and Zn from these diffuse sources in addition to wastewater from the Umicore smelter in Overpelt (Belgium).

In the Run River during rainy periods in winter, the Zn originates from the anthropogenic enriched soils via shallow groundwater alimentation with positive δ^{66}Zn isotope values. During drier summer, the Zn content in the Run decreases and the positive δ^{66}Zn values become more negative. This correlates with decreasing Zn concentrations in the Run River (Figure 4.6). Groundwater shows similar trends as the Run, thus indicating a dilution effect by deep groundwater.

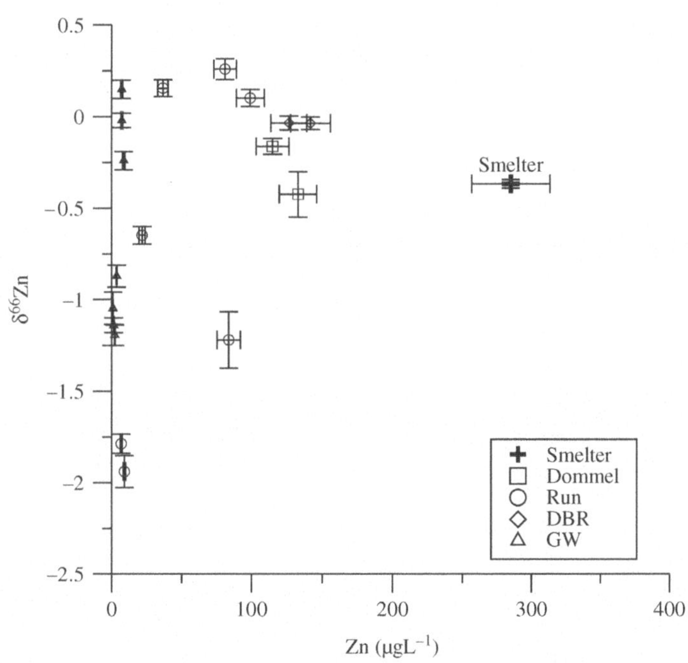

Figure 4.6 δ^{66}Zn vs. Zn concentrations in surface- and groundwater from the Dommel catchment.

Suspended matter of the Meuse River

In order to predict the contaminant dynamics in periodically flooded soils the contaminant load on the floodplains of a river catchment was first assessed. Dike breach ponds provided information about areas with dominance of sediment deposition as one important end member. Depth profiles of anthropogenic elements clearly illustrate the pollution history of the river Meuse with more recent levels showing a decrease in anthropogenic contribution for the last decades (Bakker *et al.*, to be submitted).

The above results were illustrated with the cadmium data obtained for recent suspended matter (SPM: 2001–2002 and 2005–2006) and SPM of the 1993 flood and sediments sampled in a dike pond (Figure 4.7). The sediments in the pond show a rapid decrease in Cd pollution from 1950 to 1993 and a much slower decrease in Cd from 1993 to 2002.The SPM of the 1993 flood had Cd concentrations of about 10 mg/kg and one day before the peak of the flood the cadmium decreased due to dilution of the SPM with clean sands.

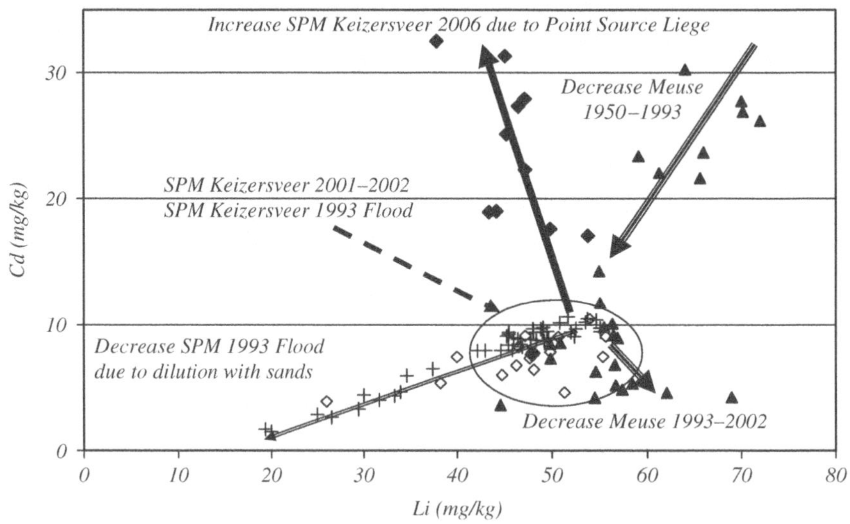

Figure 4.7 Co-variation of Cd and Li concentration for SPM in Keizersveer (2001–2002 (open diamonds); 2005–2006, solid diamonds), the Ammerzoden core (solid triangles) and the 1993 flood (crosses).

Based on the dike pond core, we predict future Cd concentrations in the SPM of Keizeersveer will to steadily decrease below 5 mg/kg. However, recent input from a point source in Liege may raise anthropogenic metal concentrations higher than observed during the maximum of the Meuse pollution in the 1950's.

Sediment dynamics and Geoacoustic methods

The development of non-invasive, acoustic sensing techniques to accurately and continuously monitor sediment transport complemented these studies to assess associated potential pollutant transport. This provides an important base to incorporate these data into hydrological models for operational water management. In this context, a novel passive acoustic system was developed and tested under a variety of well-controlled SPM flux conditions. The acoustic emissions due to inter-particle interactions – recorded near the bed with a very high level of detail – were shown to serve as an effective proxy for better understanding and quantifying sediment transport (Hermand and Verbanck, in preparation). The high sensitivity and wide dynamic range of the acoustic measurements enabled to resolve individual flood events and infer average properties of the sediment-laden flow on multiple scales under all regimes.

This included a novel combination of thorough data analysis to show statistically significant correlation between time-dependent acoustic energy and flow, flux and boundary conditions including current, turbulence, bubble collapse, suspended and bed loads, particle granulometry, surface wave and bedform configurations. First time series of particle fluxes and possibly size distribution can be obtained and events such as vortex-induced sediment bursts or the passage of coarser particles could be identified. This work addresses the question about particle re-suspension particularly during high-flow events. It allows first speculations to which bedforms control the highest particle fluxes. This is associated with further developments as to if observed highest fluxes correspond to maximal turbulence intensity. Further uncertainties include highest re-suspension efficiency of a sediment-laden flow, water-surface breaking as inhibiting factor. Such new questions are of great importance to the pollutant transport issues (Figure 4.8).

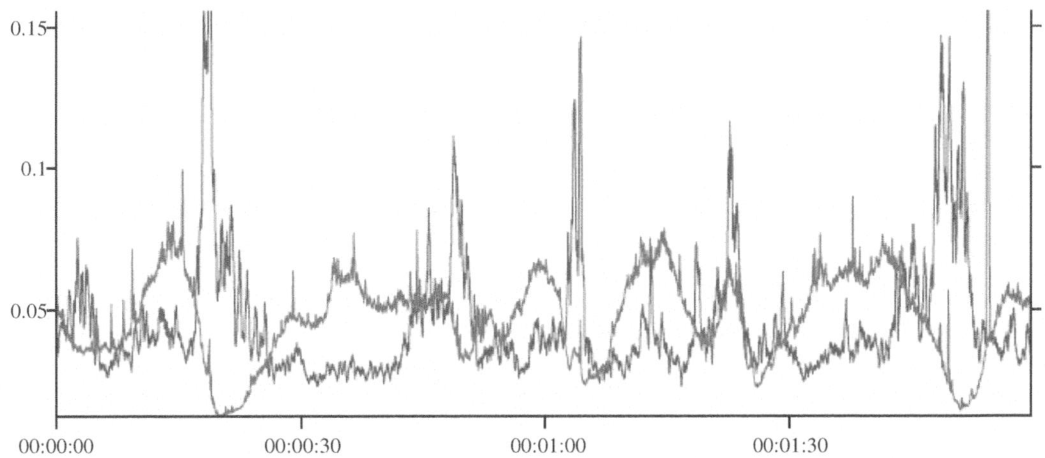

Figure 4.8 Two-minute passive-acoustics experimental record. The green curve shows the modulation of sediment flux at LeCoq site. Turbulence noise (as reflected by the blue line) allows to interpret the data records very consistently, notably in terms of surface gravity waves cycles characteristic of this type of high-streampower sediment-laden flow.

In addition variational inversion schemes of acoustic propagation data were validated for a dynamic model of bed deposits in the Scheldt estuary as constructed from acoustic reflection/backscatter measurements at the *St. Anna* site (Hermand *et al.*, 2006a, b). The scheme constitutes an effective tool of geoacoustic characterization in river and estuary and also a framework that enables the assimilation of conventional water and SPM/turbidity suspended load measurements into a sediment transport model such as the one validated for the Elbe river (Figure 4.9).

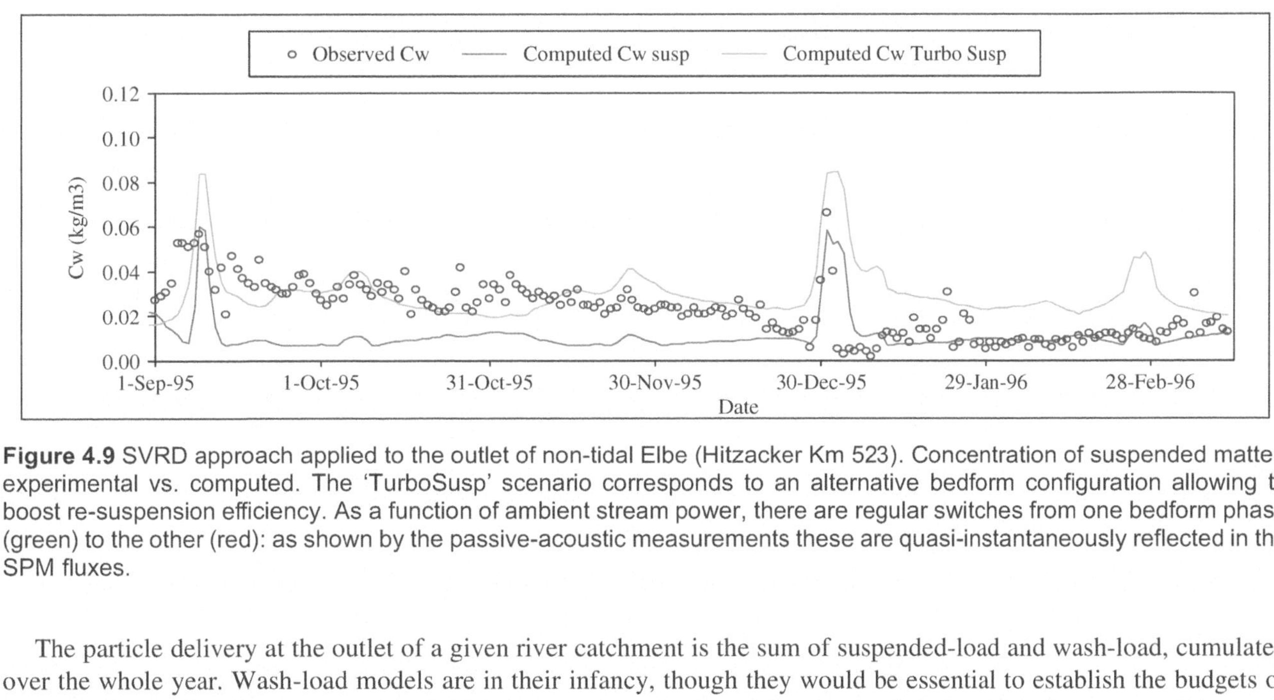

Figure 4.9 SVRD approach applied to the outlet of non-tidal Elbe (Hitzacker Km 523). Concentration of suspended matter: experimental vs. computed. The 'TurboSusp' scenario corresponds to an alternative bedform configuration allowing to boost re-suspension efficiency. As a function of ambient stream power, there are regular switches from one bedform phase (green) to the other (red): as shown by the passive-acoustic measurements these are quasi-instantaneously reflected in the SPM fluxes.

The particle delivery at the outlet of a given river catchment is the sum of suspended-load and wash-load, cumulated over the whole year. Wash-load models are in their infancy, though they would be essential to establish the budgets of some solid-bound pollutants, associated with the very-fine sized particles exported during high-flow events (Verbanck *et al.*, 2007). The outlet of non-tidal Elbe (Hitzacker, Km 523, just upstream of Geesthacht weir) was selected to test the suspended-load computation procedure developed within FLUX 3. It conclusively showed that bedform configuration changes control the importance of particle delivery fluxes (Verbanck, 2008). It was also shown, based on a systematic application of control factor m and SVRD, that a fully-coupled 1D model, accounting for the dynamic sediment-alluvial-flow adaptation, was feasible (Huybrechts *et al.*, 2008). This outlines that without the fully-coupled approach, tracking of the bed evolution would be insufficient. There is no other example of this in the literature.

CONTRIBUTIONS AND POLICY RECOMMENDATIONS FOR RIVER BASIN MANAGEMENT

There is no direct contribution of the subproject FLUX to policy but we can notice that:

FLUX 1 results indicate that most of the incoming POPs remain in the catchments and accumulate in the top-soils, with re-volatilization and degradation in the soils being very limited. For river basin managers, this means that soil pollution and the exposure of humans to POPs via the agricultural food-chain are the issues to focus on. Our research also shows that groundwaters are more or less protected from apolar POPs such as PAHs while more soluble compounds can affect ground and surface water systems such as shown for the Brévilles case.

FLUX 2 results indicate that contaminants from GW are significantly degraded at the interface just before a potential release in the River system. Furthermore, results from FLUX 2 and others showed that (organic) potential contamination transfer from GW to SW was negligible with respect to contamination coming from the River flow. Finally, isotopic study enables to gain knowledge in water origin and water transfer. However, this improvement of understanding of the phenomenon often leads to make the system more complex than initially anticipated. Therefore, it showed that the use of traditional characterisation methods can lead to a simplistic characterisation of the system. This gives rise to warnings not to implement inadequate management measure due to oversimplification. Monitoring using isotopic as ubiquitous tracers that are already in the system with analyses of low flow and high flow periods is recommended to improve knowledge of the groundwater and surface water connection. River basin management issues address by modelling of contaminants at the interface between groundwater and surface water deals with issues such as *Contaminated sediments* and *Large scale pollution of industrial sites*. The Knowledge gained in FLUX 2 about contaminated sediments and Large scale pollution of industrial sites deals with understanding of contaminant migration at the groundwater and surface water interface.

FLUX 3 results indicate that monitoring networks have to be updated regularly using state of the art geochemical tracers that combine geochemical and isotope methods. This has to be done on carefully selected sampling points together with a geochemical picture of the river during low and high discharge. Floods have also to be included in the monitoring program because they are responsible for the definitive deposition of sediments in a river basin. Moreover, most of the transported material (particulate matter and dissolved load) are exported during flood events occurring only during a few days per year. Suspended matter also needs to be monitored qualitatively because it is not possible to deduce the quality of the suspended matter from the total and dissolved monitoring results. This omission in the EU Water Framework Directive should be adjusted. Interlaboratory differences can make it impossible to interpret monitoring results, therefore the measurements for the monitoring program in a river basin should be done by only one laboratory or interlaboratory comparisons with common procedures should apply. The Ebro case study highlights the need of common multidisciplinary investigations to investigation of all the aspects of the human activities in populated and developed watersheds. This includes water regulation, data on dams, agriculture with chemical inputs and irrigation as well as urbanization and industrialization with the related wastes. Such measures can help to weigh natural phenomena versus anthropogenic ones and thus to make recommendations for river basin managers. For instance, the drastic decrease of suspended matter transported by the Ebro due to dams and the water abstraction in the lower part of the basin for irrigation induce new erosion of the river mouth and an evolution towards a quasi-permanent salt-wedge estuary. However, corrective actions are not evident, as the suspended matter exportations are controlled by the dams, regulate the water flow all year around.

OPEN & NEW QUESTIONS FROM FLUX

Although it is clear that most POPs reside in the soils in the long term, there are still questions regarding the relative importance of sub-surface transport and runoff compared with other loss processes from soils such as volatilisation and degradation.

REFERENCES

Bakker, I., Klaver, G., van Os, B., Joziasse, J., Rijnaarts, H., Négrel, Ph., Petelet-Giraud, E., to be submitted. Temporal and spatial trends of suspended matter pollution in the Meuse river.

Brenot, A., Baran, N., Petelet-Giraud, E., Négrel, Ph. (2008) Interaction between different water bodies in a small catchment in the Paris basin (Brévilles, France): Tracing of multiple Sr sources through Sr isotopes coupled with Mg/Sr and Ca/Sr ratios. *Applied Geochemistry* **23**, 58–75.

Brenot, A., Négrel, Ph., Petelet-Giraud, E., Millot, R., Roy, S. submitted. Decomposition of dissolved elements fluxes on the Ebro River Basin (Spain) supported by sulfate isotopes. Submitted to The Science of the Total Environment.

Hermand, J.P., Verbanck, M.A., In preparation. Periodicities and bedform-driven modulation of sediment transport revealed by a novel acoustic sensing technique. *J. Acoust. Soc. Amer.*

Hermand, J.P., Meyer, M., Asch, M. and Berrada, M. (2006a) Adjoint-based acoustic inversion for the physical characterization of a shallow water environment. *J. Acoust. Soc. Amer.* **119**, 3860–3871.

Hermand, J.P., Perichon, L. and Verbanck, M.A. (2006b) Characterization of sediment dynamics in an estuary environment using acoustic techniques. *J. Acoust. Soc. Amer.* **120**, 3356: 5aUW10.

Huybrechts, N., Zhang, Y.F. and Verbanck, M.A. (2008) Fully-coupled 1D model of mobile-bed alluvial hydraulics for the study of the effects of silt-removal flushes performed annually in the Lower Yellow River. Proc IAHR RiverFlow2008 Conf, Izmir.

Klaver, G., van Os, B., Negrel, Ph. and Petelet-Giraud, E. (2007) Influence of hydropower dams on the composition of the suspended and riverbank sediments in the Danube. *Environmental Pollution* **148**(3), 718–728.

Petelet-Giraud, E., Négrel, P., Gourcy, L., Schmidt, C. and Schirmer, M. (2007). Geochemical and isotopic constraints on groundwater-surface water interactions in a highly anthropized site. The Wolfen/Bitterfeld megasite (Mulde subcatchment, Germany). *Environmental Pollution* **148**(3), 707–717.

Petelet-Giraud, E., Klaver, G. and Négrel, Ph. (2009) Natural versus anthropogenic sources in the Dommel surface and groundwater dissolved load (Meuse basin): Constraints by boron and strontium isotopes and gadolinium anomaly. *Journal of Hydrology* **369**, 336–349.

Verbanck, M.A., Larcy, A., Huybrechts, N. and Vanderborght, J.P. (2007) Suction-Vortices Resupension Dynamics applied to the computation of fine-particle river fluxes, In: Sediment Dynamics and Pollutant Mobility in Rivers: An Interdisciplinary Approach, Springer, ed., Hamburg, 157–167.

Verbanck, M.A. (2008) How fast can a river flow over alluvium? *Journal of Hydraulic Research* **46**.

Chapter 5

MONITOR – Data analysis, monitoring and analytical protocols, and sampling devices

M. Petrovic, S. Meijer, K. C. Jones and D. Barceló

KEY OBJECTIVES

The subproject MONITOR was organised in three different branches with the following objectives:

MONITOR 1

- Collection of data on likely usage/emission/occurrence of the potential target compounds in the study basin/catchments
- Screening of target compounds (priority pollutants) in soil, water, sediment and biota compartments (samples from BASIN cases)
- Identification of emerging compounds and relevant metabolites in environmental samples from the BASIN cases
- Identification of degradation products for emerging contaminants

MONITOR 2

- To evaluate the available analytical methodology for WFD priority substances with the objective to harmonize protocols used in monitoring studies
- To develop and optimise advanced analytical protocols for the analysis of several groups of target compounds (selected pesticides and their metabolites, pharmaceutically active compounds, new flame retardants)
- To provide analytical protocols to be used by partners in IP for chemical analysis of target compounds in soil/water/sediment/suspended matter samples

MONITOR 3

- Validate, optimise and deploy time-integrating passive sampling devices for monitoring air concentrations, atmospheric deposition and water quality parameters
- Integration of ultrafast magnetic proxy screening into site characterization
- Optimise and validate neural networks and sampling strategies for early identification of trends from highly variable data sets.
- Investigate factors controlling the soil-air exchange.

KEY ACHIEVEMENTS

Optimised method for the analysis of pesticides in water and soil

This protocol enables the quantification of atrazine (AT), deethyl- and deisopropyl-AT, isoproturon (IPU), chlortoluron, acetochlor (ACT), acetochlor ethanesulfonic (ACT-ESA) and ACT oxanilic acid (ACT-OA) and their derivatives simultaneously in water samples. It has been tested on water media (at the Brévilles sub-catchment) but is also applicable to soil and sediment matrix. The analytical protocol is based on solid phase extraction (SPE) and liquid chromatography – tandem mass spectrometry (LC-MS-MS) and it gives reliable measurements of pesticides and its derivatives at low ng/L level.

Liquid chromatography-quadrupole-linear ion trap mass spectrometry (LC-QqLIT MS) method for the analysis of pharmaceutically active compounds in surface and wastewater

A method for the simultaneous determination of 72 multi-class pharmaceuticals was developed using off line solid phase extraction (SPE) followed by liquid chromatography – a hybrid triple quadrupole-linear ion trap mass spectrometer (QqLIT) (Gross *et al.*, 2009; project deliverable M2.10, see www.eu-aquaterra.de). In order to obtain an extra tool for confirmation and identification of the studied pharmaceuticals, and Information Dependent Acquisition (IDA) experiment was performed. Compound identification was carried out by library search with a developed MS/MS library, created by the infusion of standards, based on EPI spectra at the three collision energies. The main advantages of the developed method besides high sensitivity (limits of detection ranging from 0.1–55 ng/L, depending on the matrix), selectivity and reliability of results, is that all compounds are extracted in a single step, speeding up considerably sample preparation. Recoveries obtained were generally higher than 50% for both surface and wastewaters.

LC-QqLIT-MS method for simultaneous determination of new brominated flame retardants

The method was use to determine hexabromocyclododecane (HBCD) diastereoisomers and tetrabromobisphenol A (TBBPA) and its derivates (Guerra *et al.*, 2008; project deliverable M2.7).

The method combines two instrumental approaches: one is based on a selected reaction monitoring (SRM) method (see Figure 5.1), in which the two most abundant transitions were selected for each analyte, as well as for internal standards and the second one, the ion trap was used for storage and subsequent fragmentation of precursor ions, obtaining an enhanced product ion (EPI) experiment.

The developed methods display excellent detection limits in SRM mode (0.1–1.8 pg), but slightly better results are obtained in EPI mode (0.01–0.5 pg). Moreover, the selectivity was enhanced working with IDA performance, obtaining the complete spectrum for the fragmentation of the molecular ion, whish opens the possibility to the identification of potential degradation products of selected ions.

Deployment and validation of two types of time-integrating passive sampling devices

For atmospheric deposition, a bulk wet and dry deposition sampling device was used, consisting of a funnel and cartridge system (Figure 5.2a). For air sampling a novel passive sampling methodology, using polyurethane foam (PUF) discs was used (Figure 5.2b).

Bulk deposition samplers are used to measure deposition of POPs from the air to the soil while passive air samplers determine the concentration of these substances in the air. The two different types of passive samplers were deployed in several river catchments at coinciding sites. Results show that for some compounds data pairs follow a linear relationship between the results of the different sampler types, but this is not the case for all compounds. This indicates that meteorological factors such as temperature, wind, precipitation and also the concentration and nature of particles in the air can influence the relationship between concentrations of POPs in the air and their deposition to soil. To further validate the passive air samplers, performance reference compounds (PRCs) were used to investigate the influence of differences in wind speed at the studied sites. From the loss of these compounds site, specific uptake rates were determined, which in turn were used to calculate concentrations of POPs in air. The results show evidence that differences in PRC loss rates as well as accumulated amounts of POPs depend on differences in the average wind speed. Therefore PRCs can help to identify whether different amounts of POPs accumulated by passive air samplers are really the result of concentration differences or rather of different uptake rates.

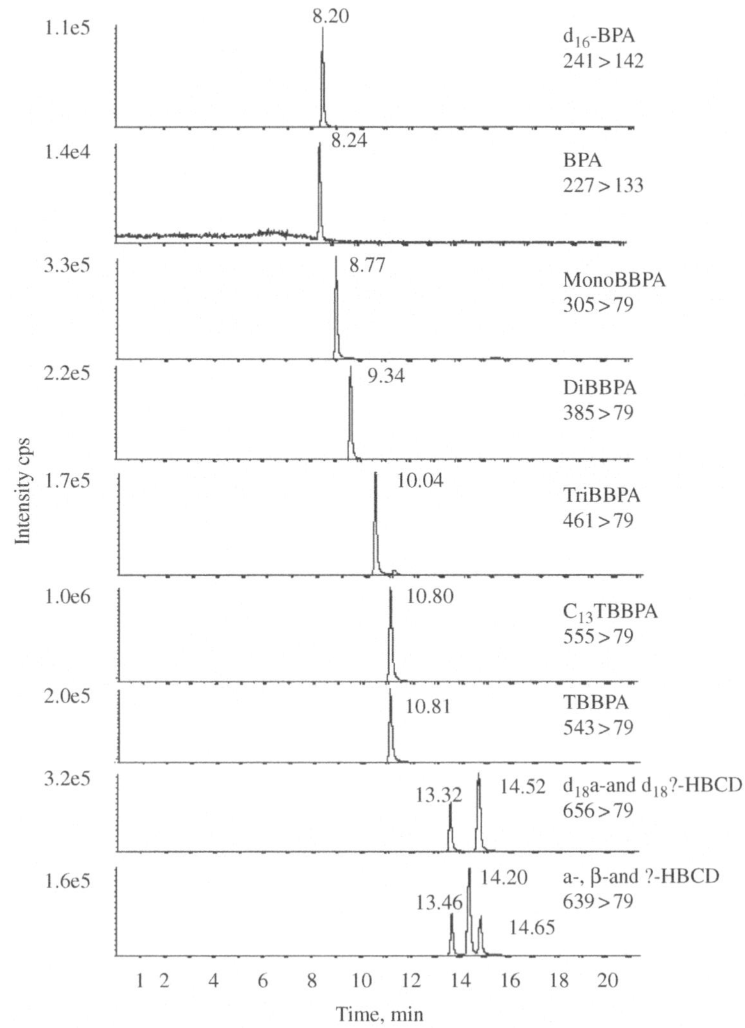

Figure 5.1 Selection reaction monitoring (SRM) chromatogram of a standard mixture.

Figure 5.2 Sampling devices for (a) atmospheric deposition and (b) air sampling.

The two types of passive samplers were successfully validated for measuring polybrominated diphenylethers (PBDEs), a compound group that is widely used as flame-retardants in many types of consumer products. Moreover, it was shown that different groups of POPs such as PCBs, PAHs, and PBDEs can be determined within one sampler using sampling periods of about two months. Hence, the sampling system is a very powerful, cost-effective tool to monitor air concentration and fluxes of POPs from the atmosphere to the soil.

Integration of ultrafast magnetic proxy screening into site characterization

To recognize local variations and to allow a targeted, cost-efficient spatial monitoring, a stepwise procedure involving simple and fast proxy methods and subsequent sampling and testing with standard methods was proposed. Magnetic proxy screening was used to outline local variations of ambient conditions as well as of relative pollution loads within two selected test areas in the Danube and Elbe Basin cases. The resulting three dimensional maps of relative distributions of magnetic parameters serve as a proxy for industrial pollution and allow a better targeted selection of sampling and testing locations.

Magnetic proxy screening has been used to investigate two areas in the river Basins of Danube and Elbe. In all test areas, magnetic proxy screening seems to be able to delineate contaminated areas at different scales.

Optimisation and validation of artificial neural networks to allow early identification of trends from highly variable data sets

Neural network methods were used to analyse the data obtained within this workpackage, with the aim to optimize sampling strategies. Artificial neural networks (ANN) were implemented in the MATLAB programming language and were used to identify trends from these highly variable data sets. Applications include the prediction of time series in the Ebro catchment, the interpolation of spatially distributed magnetic susceptibility data (Danube and Elbe catchments) and the cluster analysis of spatially distributed sampling data of pollutants (all catchments). These calculations allow an estimation of the number of locations and time intervals between measurements and other measurement parameters needed for an optimal sampling strategy for environmental monitoring.

IMPLICATIONS AND APPLICATION OF RESULTS

The analytical protocols developed enable to quantify emerging compounds (pesticides, pharmaceuticals, illicit drugs, polar pesticides and new brominated flame retardants) and some of their derivatives. Quantification of these compounds in the environment will give indication of the occurrence of these compounds in the environment and may contribute to decide whether or not these compounds shall be included in the priority list defined in the context of the WFD.

The results show that passive samplers are a cost-effective and reliable method for monitoring pollution in river basins. In combination with proxy screening and/or statistical analysis methods such as artificial neural networks, the local monitoring results can be successfully scaled up to the catchment scale.

CONTRIBUTIONS AND POLICY RECOMMENDATIONS FOR RIVER BASIN MANAGEMENT

Locally/nationally

The analytical protocols developed within MONITOR address key river basin management issues such as *Diffuse pollution by agriculture* (contamination by pesticides) and *Emerging compounds* (*for example contamination by pharmaceuticals*). A systematic monitoring of selected emerging contaminants in the Ebro river basin is recommended to the river basin management in order to get more comprehensive picture of the river basin chemical status.

In a European context

Regulatory agencies have issued detailed guidelines on how pharmaceuticals should be assessed for possible unwanted effects in the environment. The first requirement was introduced by the European Union in 1995 with 92/18 EEC directive and the corresponding note for guidance for veterinary pharmaceuticals. The European Commission released a draft guideline (Directive 2001/83/EC) specifying that an authorization for a medicinal product for human use must be accompanied by an experimental risk assessment. Evaluation of the potential environmental risk posed by pharmaceuticals should be evaluated on a case-by-case basis where MONITOR contributed by providing optimized and validated analytical protocols.

MONITOR has evidenced that monitoring data of emerging contaminants are often not comparable from one river basin to another (or one country to another). Harmonization of protocols and more interlaboratory comparisons are needed

MONITOR recommends inclusion of main transformation products of selected pollutants on the list of monitored compounds

MONITOR recommends development of risk-assessment methodologies and advanced measuring systems for different stages of wastewater treatment plants (WWTP) and to use them in existing and new WWTPs to ensure good performance of WWT technologies for urban, industrial and agricultural purposes.

MONITOR recommends the use of passive samplers in combination with scaling techniques as a tool for monitoring of organic pollutants in river catchments.

OPEN & NEW QUESTIONS FROM MONITOR

It is important to incorporate metabolites excreted by humans and transformation products formed during wastewater treatment in monitoring surveys as part of a risk evaluation for environmentally relevant pharmaceuticals. Therefore, elucidation of bio(photo)degradation pathways and identification of main transformation products is crucial task.

Further optimization of the analytical methodology for emerging contaminants and inclusion of transformation products (bio and photo degradation products and metabolites) into the analytical protocols.

Further validation of passive sampling techniques is required, in particular with respect to combining techniques to optimize sampling strategies.

REFERENCES

Gros, M., Petrović, M. and Barceló, D. (2009) Tracing pharmaceutical residues of different therapeutic classes in environmental waters by using liquid chromatography/quadrupole-linear ion trap mass spectrometry and automated library searching. *Analytical Chemistry* **81**(3), 898–912.

Guerra, P., De La Torre, A., Martínez, M.A., Eljarrat, E. and Barceló, D. (2008) Identification and trace level determination of brominated flame retardants by liquid chromatography/quadrupole linear ion trap mass spectrometry. *Rapid Communications in Mass Spectrometry* **22**(7), 916–924.

Chapter 6

TREND – Temporal development of environmental media: Properties, processes and stresses

G. J. Lair, M. H. Gerzabek, C. Bürger, H.-P. Broers, C. A. M. van Gestel, P. Hsu, A. A. Koelmans, A. Poot, D. M. E. Slijkerman and A. Visser

KEY OBJECTIVES

The objective of sub-project TREND was to obtain insights into the temporal development of different compartments in the soil-groundwater-sediment-river system. This is important to decide to what extent these compartments are impacted by various perturbations with respect to their functioning. Within the sub-project the workpackage TREND 1 focussed on trends in soils and sediments, TREND 2 on groundwater and TREND 3 on modelling and forecasting from catchment to basin scales. Specific objectives were:

- To study the temporal variability of soil properties in river floodplains in relation to climate and land use to assess their effects on soil and sediment ecological functioning
- To measure the temporal variability of pollutant concentrations in floodplain soils and sediments due to inundations and to address bioavailability to soil and sediment biota
- To understand the resilience of natural cycling processes in soils and sediments towards contamination and inundation, taking organic material mineralisation as a model process
- To develop Integrated Quality Indicators (IQI), indicative for trends in ecological functioning in sediments to assist river basin management
- To develop operational methods (lumped and deterministic), which assess, quantify and extrapolate trends in groundwater water systems at various scales (minimum size is that leading to consideration as "water body" in the Groundwater Directive)
- To link changes in land use, climate and contamination history to changes in groundwater water chemistry (e.g., nutrients and pesticides). The method is directed to implementation in the Water Framework Directive and Groundwater Directives
- To develop an artificial neural network (ANN) simulator for trend analysis in river basins and for optimising monitoring strategies (e.g., evaluating the sensitivity of selected river basins to climate change scenarios)
- To promote the collaboration within the IP such that innovative methods (ANN) and skills can be applied to develop sustainable use of European water resources

©2010 IWA Publishing. *Advanced Tools and Models to Improve River Basin Management in Europe in the Context of Climate Change (AquaTerra)* by Michael Finkel, Johannes Barth and Peter Grathwohl ISBN: 9781843393726. Published by IWA Publishing, London, UK.

KEY ACHIEVEMENTS

Trends in soil and sediment

Since the Middle Ages, humans have increasingly altered the size, structure and dynamics of many European rivers and floodplains, which act as buffers for flood waters and as filter and storage sites for pollutants. Especially in the 19th and 20th century, river channelization and the construction of dams for flood protection and hydropower plants have severely changed river hydrology, sediment fluxes and floodplain erosion/sedimentation patterns. Dikes along river channels physically limit the area inundated during floods. Constructed levees stabilize the river channel, which may promote riverbed incision and reduce the frequency of floodplain inundation (Tockner and Stanford, 2002). As a consequence, the morphological age of riparian areas along anthropogenically impacted rivers has considerably increased over the last century (e.g., Lair *et al.*, 2009c). This is also reflected by increasing replacement of pioneer plant communities with later successional stages of vegetation (Tockner and Stanford, 2002). This marks a shift from wetland and semi-terrestrial towards less frequently inundated and terrestrial ecosystems. Our analyses at the Danube River show that the development towards a terrestrial ecosystem is accompanied by considerable changes in the soil environment within 600 years (Lair *et al.*, 2009d), such as rapid carbon accretion and organic matter pool stabilization (Lair *et al.*, 2007c; Zehetner *et al.*, 2009a), changes in nutrient contents (Zehetner *et al.*, 2008; Lair *et al.*, 2009b,c), increasing crystallinity of soil born iron-oxide minerals (Lair *et al.*, 2009a) and the dissolution of readily soluble minerals (Zehetner *et al.*, 2009b).

Floodplains are a sink and memory for pollution that reaches the river system via wastewaters, surface waters or atmospheric deposition. Our analyses reveal that the above-mentioned (quantified) soil changes generally increase the retention and decrease the mobility of cationic and organic pollutants as floodplains develop towards terrestrial ecosystems (Graf *et al.*, 2007; Lair *et al.*, 2008; Lair *et al.*, 2009b,c; see also Figure 6.1). However, we also find that intensive agricultural use can severely decrease the retention capacity of floodplain soils for organic contaminants and heavy metals through excessive aeration (tillage) and subsequent loss of soil organic matter (Zehetner *et al.*, 2009a; Lair *et al.*, 2009c; see also Figure 6.1). Presently, as much as 79% of the riparian area of European rivers is intensively cultivated (Tockner and Stanford, 2002). In long-term field experiments it was well demonstrated, that various farming/ manuring practices can change the soil's retention capacity for contaminants up to a factor of three (e.g., Lair *et al.*, 2006; Lair *et al.*, 2007a,b; Novoszad *et al.*, 2005).

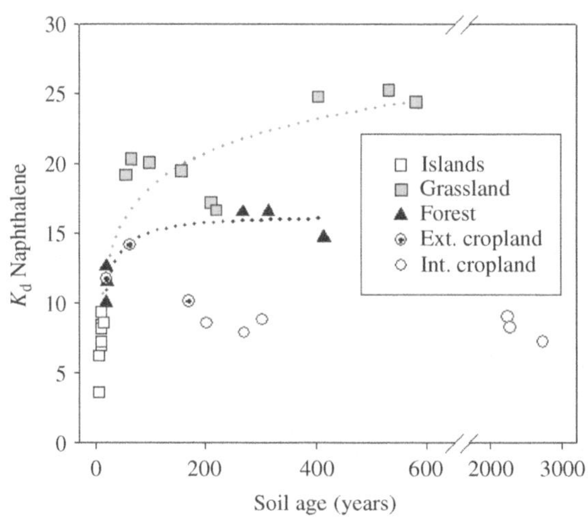

Figure 6.1 Impact of soil development and landuse on the retention of naphthalene in floodplain topsoil (0–10 cm) at the Danube River (Lair *et al.*, 2009c). K_d, Partition coefficient according to OECD (2000), an increasing K_d value reflects increasing adsorption to the soil matrix; ext., extensive; int., intensive.

Our analyses also show that temporal contaminant bioavailability and toxicity in dynamic river-floodplain systems are a result of complex interactions between nutrients, hydrology and pollutant levels (Poot *et al.*, 2007; Hsu *et al.*, 2007). In the Elbe River, for instance, algae took up contaminants from the water phase starting in April, and after dying off in June, they settled and transferred contaminants to the upper sediment layer, from where they could easily be re-mobilised (Hsu *et al.*, 2007; Figure 6.2). In July, the sediment toxicity dropped again whereas water toxicity increased considerably. This may be the result of enhanced organic matter degradation and contaminant release due to enhanced microbial activity in the sediment, which in turn might have been stimulated by the presence of fresh substrate in the form of dead algal cells.

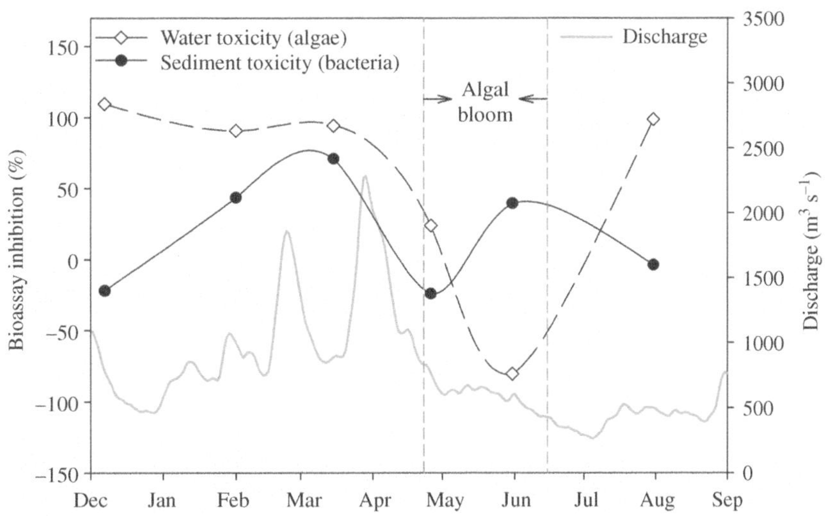

Figure 6.2 Seasonal variation in ecotoxicity of river water (shown with algal inhibition) and upper layer sediments (shown with bulk sediment bacteria inhibition) in relation to water discharge and algal bloom in the Elbe River in 2005.

High food quantity and quality in floodplain soils and sediments caused by high fluxes of nutrients (e.g., nitrogen, phosphorous) frequently mask adverse effects of recent and stored sediment contamination. While ecotoxicity can be an important integral indicator for contaminant loadings, the vitality of soil organisms is often controlled by other factors such as pH, organic matter, clay and nutrient contents (Slijkerman, 2006; Bleeker and Van Gestel, 2007; see Figure 6.3).

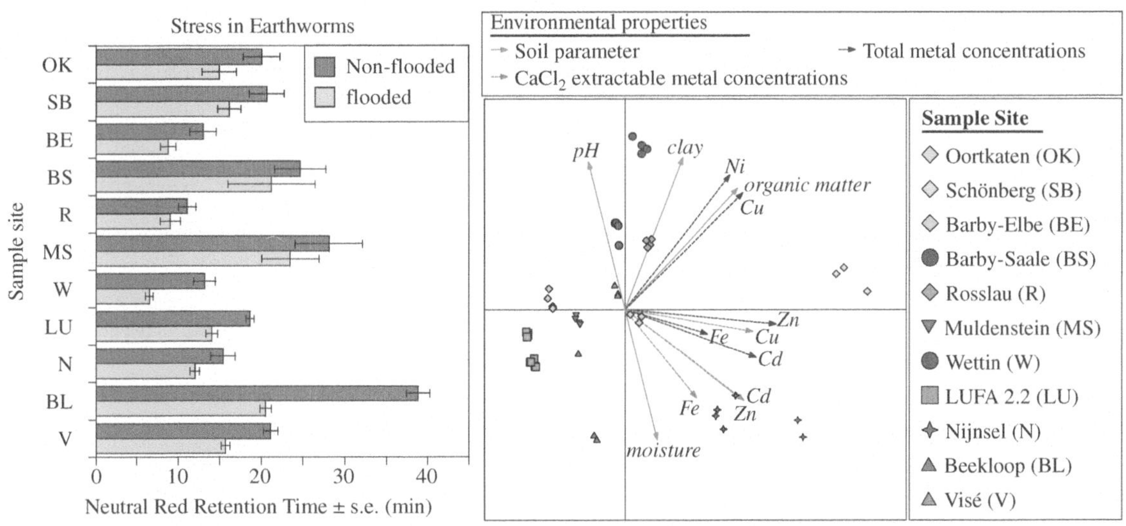

Figure 6.3 Stress in earthworms exposed in bioassays to floodplain soils (left; high NRRT reflects low stress on earthworm species). Principal Component Analysis shows the importance of non-metal properties in explaining differences between floodplain soils.

Most important findings of TREND1 and some invited authors are summarized in a joint review article (Lair *et al.*, 2009c) describing (1) the evolution and dynamics of floodplains, (2) the input and retention of contaminants in floodplain soils and sediments, (3) soil development and its impact on contaminant retention, (4) storage and fate of contaminants (nutrients, heavy metals, organic pollutants) in dynamic river-floodplain systems, and (5) the ecotoxicity of contaminants in polluted river-floodplain systems.

Trends in groundwater

Several catchments were selected for TREND 2 research (Table 6.1), with different sizes and geology (project deliverable T2.1, see www.eu-aquaterra.de). For each catchment, land use and climate was reconstructed (project deliverable T2.2) to provide the input function of contaminants to the groundwater system (Brouyere *et al.*, 2004b). Similar trends of nitrate loads were observed, increasing since the 1950s, but often decreasing after the 1980s.

Table 6.1 Case studies in TREND 2

Sub-basin	Hydrogeological characteristics	Spatial scale
Dommel/Brabant	Unconsolidated Plesitocence deposits; fine to medium coarse sands, loam	500/5000 km^2
Wallony-Hesbaye	Cretaceous chalk, fissured, dual porosity aquifer	440 km^2
Wallony-Pays de Herve	Cretaceous chalk and sands, fissured	285 km^2
Wallony-Néblon	Carboniferous limestone, folded karstified	65 km^2
Wallony-Meuse alluvial plain	Unconsolidated deposits; gravels, sands and clays	125 km^2
Brévilles	Lutecian limestone over Cuise sands, limestone fissured	2.5 km^2

Two approaches for the detection of trends in groundwater quality were deployed: statistical methods and physically-deterministic modelling (see also Table 6.2). For each catchment, the relation between the concentrations of contaminants on one hand and the depth or time of sampling on the other hand was determined (project deliverable T2.3). Groundwater dating – using $^3H/^3He$, CFCs and/or SF_6 – provided insight in the distribution of groundwater age in each catchment. Trends in groundwater quality were determined using statistical trend analysis (Batlle Aguilar *et al.*, 2007), and age-scaling (Visser *et al.*, 2007). These trends were extrapolated to predict future changes in contaminant concentrations.

Physically-deterministic groundwater flow and contaminant transport were build for each catchment or selected areas within the catchment (project deliverables T2.5, 2.7, 2.8). These models were used to determine trends in groundwater quality, as well as extrapolate these trends into the future (project deliverable T2.9).

Modelling and forecasting from catchment to basin scales

Literature studies using traditional artificial neural networks (ANNs) for data-based modelling revealed two drawbacks (project deliverables T3.1 and T3.3): (a) the determination of the ideal network structure requires a computation-time consuming procedure. (b) The outcome of the ANN training procedure may not be the optimal training solution. This means that a better solution may exist that is not found within the training procedure. In search for better alternatives two new algorithms were discovered, investigated and applied for rainfall-runoff modelling to the Upper Gallego catchment in the Ebro river basin (Bürger *et al.*, 2007). In order to illustrate this application consider the data displayed on the upper three diagrams on Figure 6.4. These time series (air temperature, precipitation, reservoir levels) constitute the input of a function (to be estimated) whose output (discharge at river gauge Anzanigo) is the time series displayed in the lowermost diagram (Figure 6.4).

The comparison of the performance of the estimated functions on a so-called test data set (i.e., data that was not used within the procedure vs. the data set that was used, the so-called training data set, cf. project deliverable T3.1) yielded the best overall performance for the relevance vector machines (RVM) (Figure 6.5).

Another important feature of the RVM is that it does not only predict certain function values, but also provides confidence bounds for these values. This should make the RVM algorithm the method of choice for data-based rainfall-runoff modelling, where uncertainties are frequently encountered.

Regarding the second objective of TREND 3, a survey of AquaTerra databases was carried out (project deliverable T3.4) in search of additional applications of new algorithms. Using a new so-called bootstrapping technique model was successfully developed to predict the soil sorption behaviour in soils at the River Danube (project deliverable T1.22).

IMPLICATIONS AND APPLICATION OF RESULTS

We verified that soil chronosequences on riverine floodplains and long-term field experiments are useful tools to quantify changes in soil properties in time scale and under various management practices. The linkage of iron oxide crystallinity (ratio of oxalate- to dithionite extractable iron) to Cs-137 distribution and optically luminescence data allowed revealing episodes of sediment deposition and soil formation in young riverine floodplains. This knowledge can be used for further approaches in soil science, hydrology and archaeology.

Recently, the application of the traceability concept on ecotoxicological studies has been described (Ahlf and Heise, 2007). This means that ecological tools are applicable for monitoring purposes and TREND 1 data demonstrated clearly, that the seasonal variation of sediment quality has to be considered in monitoring programmes. Ecotoxic effects shall be monitored at least once a month in order to assess seasonal variabilities and to determine a typical ecotoxicological pattern. Long-term trends can only be established through comparison of results over a few years' records.

Table 6.2 Summary table comparing the strengths and weaknesses of each of the trends analysis methodologies deployed within the TREND 2 module.

	Purely statistical approaches	**Transfer function approaches**	**Age dating**	**Deterministic modelling with poor fit to the data**	**Deterministic modelling with good fit to the data**
Prerequisite	Collection of monitoring data in the field	Collection of monitoring data in the field + Collection of information on the input flux (rainfall, and either inputs or land use)	Collection of monitoring data in the field + information on the evolution of the input function + analysis of tracers in samples	Collection of monitoring data in the field + Heavy effort in collection of additional information (other piezometers, pumping and tracer tests, geophysics, soil mapping)	Collection of monitoring data in the field + Heavy effort in collection of additional information (other piezometers, pumping and tracer tests, geophysics, soil mapping)
Associated cost magnitude (on top of the data collection effort)	None	1 (surveys if not already available – purchasing of met data)	10	100 (geophysics, additional piezometers, soil mapping)	100 (geophysics, additional piezometers, soil mapping)
Understanding of the system?	No understanding of the system	Functional understanding of the system (identification of the key factors and understanding of their influence)	Functional understanding of the system	Detailed data on the system, but lack of overall understanding of the functioning of the system (exemplified by the lack of fit of the deterministic model)	Potential detailed understanding of the system under study
Extrapolation potential	Poor	Good	Good	Poor	Good
Potential universality to all systems	Potentially	Potentially	Limited only applies to homogeneous systems	Potentially	Potentially
Potential for operational use	–	++	+	–	+
Knowledge about the functioning of the system	–	++	+	++	+++

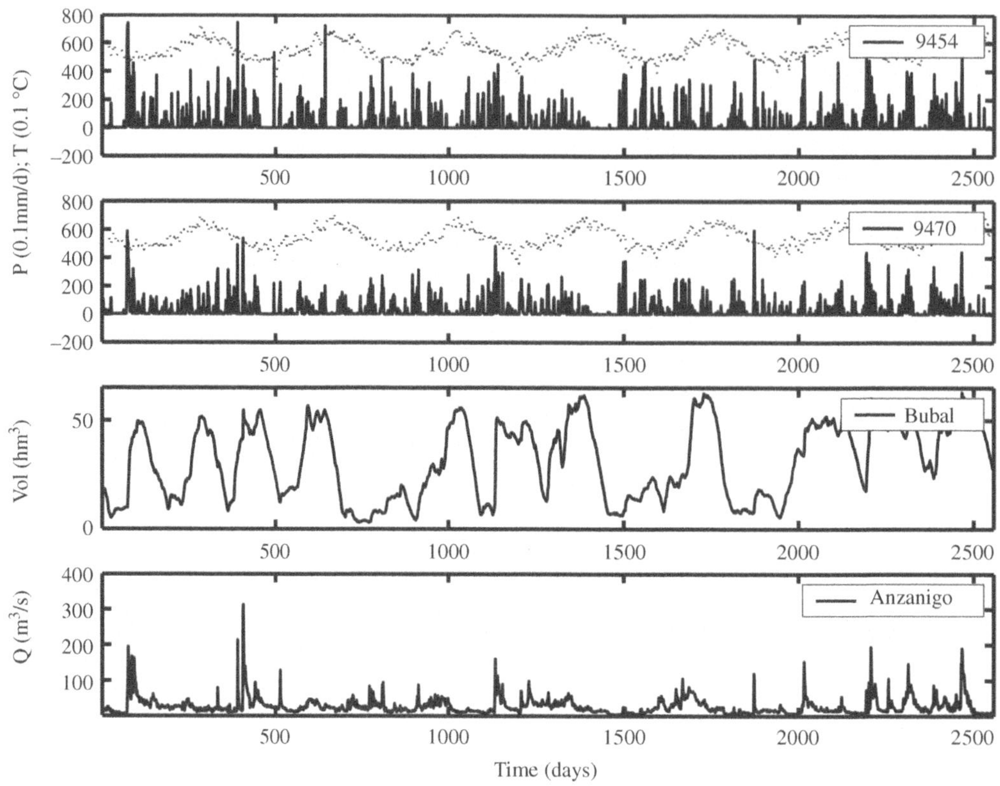

Figure 6.4 Data series for the selected study period from 1981/82 to 1987/88. Temperature values (T) are stippled, whereas precipitation (P), reservoir volumes (Vol) and discharge (Q) are drawn as solid lines. Note that the temperature series have been shifted by 450 × 0.1°C for better visualization.

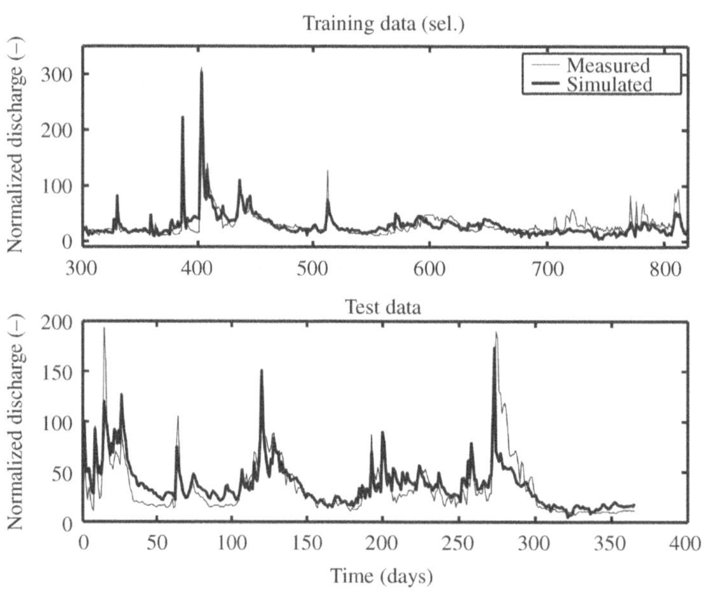

Figure 6.5 Best RVM model on training and test data.

AVS-SEM is a key parameter to consider for monitoring metal availability in sediments. In the USA, sediment quality criteria already account for AVS (US-EPA, 2005). Recently the AVS–SEM concept has also been introduced as a correction for bioavailability in a draft report on the risk assessment of zinc, which was written within the framework of Council Regulation 793/93/EEC on Existing Chemicals (Bodar *et al.*, 2005).

Trend detection in groundwater is one of the key aspects of the new Groundwater Directive and results needs to be included in the river basin management plans. Important outcomes of TREND2 are that it is necessary to aggregate trend results over the groundwater body scale in order to get useful results and to reduce the overall effects of aquifer and flow heterogeneity. An assessment of the hydrologic response times of the groundwater systems is key to the understanding and detection of trends (Broers, 2004; Visser *et al.*, 2007). Groundwater bodies are a store of contaminants which may be released to surface water over a very long time scale, long after the contamination has stopped. Contamination of groundwater systems with very short residence times will quickly have a large effect on surface water quality and the ecological state of the system. The impact can also quickly be reduced by reducing the contamination. Contamination of groundwater systems with a very long residence time will have little direct effect on surface waters. But this effect cannot be stopped by reducing the contamination and penetration of deeper parts of the groundwater systems may threaten drinking water supply for future generations.

Based on the RVM Upper Gallego catchment simulator forecasts for future daily runoff values can be made, if future precipitation, temperature (e.g., from future climate change scenarios) and reservoir time series are available. Recently, it appears that an extra session on Hydroinformatics at the EGU conference is becoming permanently established. The algorithms and statistical techniques investigated within TREND 3 are of immediate interest to this community. Furthermore, these techniques could be useful for neighbouring disciplines like agronomy, irrigation hydrology or atmospheric sciences.

CONTRIBUTIONS AND POLICY RECOMMENDATIONS FOR RIVER BASIN MANAGEMENT

Anthropogenic alterations of and increased pollutant deposition in floodplains have become a reality in most river systems. Future anthropogenic interventions affecting floodplain hydrology, such as floodplain restoration projects, need to consider the risk of mobilization of stored pollutants. Future riparian land use and management planning need to aim at soil organic matter conservation, which facilitates degradation and long-term retention of pollutants.

The vitality of soil and water organisms is strongly controlled by pH, organic matter, clay and nutrient contents (e.g., nitrogen, phosphorus) than by contamination levels. Therefore, these soil and sediment factors are to watch closely in floodplain management planning.

A close cooperation between regional and national (ground)water authorities was established in most of the case studies (Geer, Dommel and Brevilles). The methods developed in TREND 2 have become part of the trend analysis procedure in the Dutch Meuse basin and are probably used in the Dutch national procedures for the implementation of the WFD and Groundwater Directive. The implementation of the EU Water Framework Directive (EU, 2000) and the Groundwater Directive (EU, 2006) ask for specific methods to detect the presence of long-term anthropogenically induced upward trends in the concentration of pollutants in groundwater. Specific goals for trend detection have been under discussion during the preparation of the Groundwater Directive. The Directive defines criteria for the identification and reversal of significant and sustained upward trends and for the definition of starting points for trend reversal. TREND 2 has specifically provided the tools to detect trends and trend reversal of groundwater quality (see Figure 6.6). These methods and concepts have been introduced in EU Guidances for the WFD and GWD (for example: EU, 2008) and in the Common Implementations Strategy of the WFD through presentations in EU Working Group C and contributions to book chapters (for example Broers, 2008; Broers *et al.*, 2009).

River basins are very complex systems, so that extended data are necessary to study such systems in detail. From a scientific viewpoint, it is recommended to measure and collect hydro(geo)logic variables continually and to make the data available to the scientific community. Data shall be structured by the natural system boundaries rather than political ones and be offered in an easy-processing format (i.e., digital, matrix-based arrangements in ASCII formats rather than tables in pdf-files). We regret that EU states do not perform a sound assessment of programmes of measures and prioritize them in the river basin management plans (RBMPs).

OPEN & NEW QUESTIONS FROM TREND

We quantified organic carbon accretion and examined changes of organic matter pools under different land use across an age gradient in a floodplain at the River Danube. Still, carbon dynamics in river ecosystems are poorly understood and rarely quantified. Information on carbon accretion and loss in riverine floodplains under different climates and within short time frames (i.e., decades to centuries) would help to better balance carbon cycle and carbon transport to the ocean.

Yet too little is known about the particular sorption reactions with single soil constituents and about the predominant sorption processes in floodplains, where equilibrium conditions are rarely reached and different surfaces react with each other.

We concluded that in dynamic (AVS dominated) river systems, flow velocity is a key to understand the variability in metal availability. Unfortunately, insufficient data is available to fully test this hypothesis, especially when no sulphides are present in the river system.

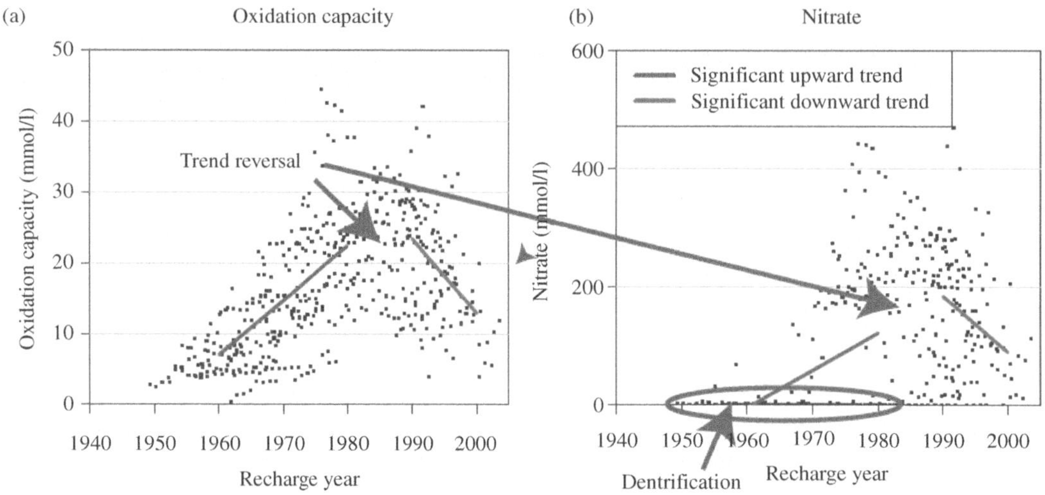

Figure 6.6 Trend reversal was demonstrated significantly (P < 0.005) for oxidation capacity and nitrate.

Furthermore, it should be verified which factors determine (bio)availability, toxicity and adaptation of biota to contaminants. Sediment toxicity could be a dynamic process during the year and it is not clear how sediment quality criteria based on equilibrium approaches could include such variability.

High Cd levels in earthworms also posed a risk for the food chain. It is still unclear how this influences the consumers of earthworms (and other species higher up in the food chain).

Future groundwater research should emphasize on the two following aspects:

(1) Groundwater and surface water interactions at short and long time scales as well as at catchment and river basin scales, including the effects of climate change on those interactions. Promising research has been published in the AquaTerra special issue of Environmental Pollution (Kalbus *et al.*, 2007; Rozemeijer and Broers, 2007). However, interactions between contaminants and the hyporeic zone are not well known. This requires collaboration in research effort from both, the groundwater and surface water research community, to study the physical, chemical and biological aspects of these interactions.

(2) Develop tools for managing groundwater resources. Future research should focus on the development of methodologies and modelling tools to predict the effects of local and regional measures, including dynamic coupling of multi-compartment models (air-soil-groundwater-marine-surface water), and to develop smart, not too expensive monitoring systems, which integrate groundwater and surface water monitoring and which preferably are linked to scenario analysis and modelling the effect of measures.

REFERENCES

Ahlf, W. and Heise, S. (2007) Quality assurance of ecotoxicological sediment analysis. In: *Sediment Dynamics and Pollutant Mobility in Rivers – An Interdisciplinary Approach.* Westrich, B. and Förstner, U. (eds.), Chapter 10.1, pp. 380–391, Springer, Berlin.

Batlle Aguilar, J., Orban, P., Dassargues, A. and Brouyère, S. (2007) Identification of groundwater quality trends in a chalk aquifer threatened by intensive agriculture in Belgium. *Hydrogeology Journal* **15**(8), 1615–1627.

Bleeker, E.A.J. and van Gestel, C.A.M. (2007) Effects of spatial and temporal variation in metal availability on earthworms in floodplain soils of the river Dommel, The Netherlands. *Environmental Pollution* **148**, 824–832.

Broers, H.P. (2004) The spatial distribution of groundwater age for different geohydrological situations in the Netherlands: implications for groundwater quality monitoring at the regional scale. *Journal of Hydrology* **299**(1–2), 84–106.

Broers, H.P. (2008) Assessing and aggregating trends in groundwater bodies. Proceedings EU Conference on Groundwater Policy developments. Good status objectives and integrated management planning. Nov 13–15th, Paris.

Broers, H.P., Visser A., Chilton, J.C. and Stuart, M.E. (2009) Chapter 4.2 Assessing and aggregating trends in groundwater quality. In: *Groundwater quality assessment and monitoring.* Quevauviller, P., Fouillac, A.M., Grath, J. and Ward, R. (eds.).

Brouyère, S., Dassargues, A. and Hallet, V. (2004b) Migration of contaminants through the unsaturated zone overlying the Hesbaye chalky aquifer in Belgium: A field investigation. *Journal of Contaminant Hydrology* **72**(1–4), 135–164.

Bürger, C.M., Kolditz, O., Fowler, H.J. and Blenkinsop, S. (2007) Future climate scenarios and rainfall-runoff modelling in the Upper Gallego catchment (Spain). *Environmental Pollution*, **148**(3), 842–854.

EU (2000) Water Framework Directive 2000/60/EC.

EU (2006) Directive 2006/118/EC on the Protection of Groundwater against Pollution and Deterioration.

EU (2008) Guidance document no. 18. Guidance on groundwater status and trend assessment. Common Implementation Strategy for the Water Frame Work Directive (2000/60/EC).

Graf, M., Lair, G.J., Zehetner, F. and Gerzabek, M.H. (2007) Geochemical fractions of copper in soil chronosequences of selected European floodplains. *Environmental Pollution* **148**, 788–796.

Hsu, P., Matthai, A., Heise, S. and Ahlf, W. (2007) Seasonal variation of sediment toxicity in the Rivers Dommel and Elbe. *Environmental Pollution* **148**, 817–823.

Kalbus, E., Schmidt, C., Bayer-Raich, M., Leschik, S., Reinstorf, F., Balcke, G.U. and Schirmer, M. (2007) New methodology to investigate potential contaminant mass fluxes at the stream-aquifer interface by combining integral pumping tests and streambed temperatures. *Environmental Pollution* **148**(3), 808–816.

Lair, G.J., Gerzabek, M.H., Haberhauer, G., Jakusch, M. and Kirchmann, H. (2006) Response of the sorption behavior of Cu, Cd, and Zn to different soil management. *Journal of Plant Nutrition and Soil Science* **169**, 60–68.

Lair, G.J., Gerzabek, M.H. and Haberhauer, G. (2007a) Sorption of heavy metals on organic and inorganic soil constituents. *Environmental Chemistry Letters* **5**, 23–27.

Lair, G.J., Gerzabek, M.H. and Haberhauer, G. (2007b) Retention of copper, cadmium and zinc in soil and its textural fractions influenced by long-term field management. *European Journal of Soil Science* **58**, 1145–1154.

Lair, G.J., Gerzabek, M.H., Fiebig, M. and Zehetner, F. (2007c) Verwitterung von Auböden in verschiedenen klimatischen Regionen Europas, Mitt. Österr. Bodenkdl. Ges. **74**, 35–52.

Lair, G.J., Graf, M., Zehetner, F. and Gerzabek, M.H. (2008) Distribution of cadmium among geochemical fractions in floodplain soils of progressing development. *Environmental Pollution* **156**, 207–214.

Lair, G.J., Zehetner, F., Hrachowitz, M., Franz, N., Maringer, F.-J. and Gerzabek, M.H. (2009a) Dating of soil layers in a young floodplain using iron oxide crystallinity. Quat. Geochronol. **4**, 260–266.

Lair, G.J., Zehetner, F., Khan, Z.H. and Gerzabek, M.H. (2009b) Phosphorus sorption-desorption in alluvial soils of a young weathering sequence at the Danube River. Geoderma **149**, 39–44.

Lair, G.J., Zehetner, F., Fiebig, M., Gerzabek, M.H., van Gestel, C.A.M., Hein, T., Hohensinner, S., Hsu, P., Jones, K.C., Jordan, G., Koelmans, A.A., Poot, A., Slijkerman, D.M.E., Totsche, K.U., Bondar-Kunze, E. and Barth, J.A.C. (2009c) How do riverine floodplains change over time and what does that mean for contaminant behaviour? An interdisciplinary analysis. *Environmental Pollution* **157**, 3336–3346.

Lair, G.J., Zehetner, F., Klepsch, S. and Gerzabek, M.H. (2009d) Schadstoffverhalten in Flüssen und Aulandschaften. GAIA 18/3, 267–269.

Novozsad, M., Gerzabek, M.H., Haberhauer, G., Jakusch, M., Lischka, H., Tunega, D. and Kirchmann, H. (2005) Sorption of naphthalene derivatives to soils from a long-term field experiment. Chemosphere **59**, 639–647.

OECD, 2000. OECD Guideline for the Testing of Chemicals 106. Adsorption – Desorption Using a Batch Equilibrium Method. OECD (Organisation for Economic Co-operation and development), Paris.

Poot, A., Gillissen, F. and Koelmans, A.A. (2007) Effects of flow regime and flooding on heavy metal availability in sediment and soil of a dynamic river system. *Environmental Pollution* **148**, 779–787.

Rozemeijer, J.C. and Broers, H.P. (2007) The groundwater contribution to surface water contamination in a region with intensive agricultural land use (Noord-Brabant, The Netherlands). *Environmental Pollution* **148**(3), 695–706.

Slijkerman, D.M.E. (2006) Application of functional endpoints in water quality assessment. Ph.D. thesis, VU University Amsterdam.

Tockner, K. and Stanford, J.A. (2002) Riverine flood plains: present state and future trends, *Environ. Conserv.* **29**, 308–330.

US-EPA, (2005) Procedures for the derivation of equilibrium partitioning sediment benchmarks (ESBs) for the protection of benthic organisms: metal mixtures cadmium, copper, lead, nickel, silver and zinc). EPA-600-R-02-011, US Environmental Protection Agency, Washington, DC.

Visser, A., Broers, H.P., van der Grift, B. and Bierkens, M.F.P. (2007) Demonstrating trend reversal of groundwater quality in relation to time of eecharge determined by 3H/3He. *Environmental Pollution* **148**(3), 797–807.

Zehetner, F., Lair, G.J., Maringer, F.-J., Gerzabek, M.H. and Hein, T. (2008) From sediment to soil: Floodplain phosphorus transformations at the Danube River. *Biogeochemistry* **149**, 39–44.

Zehetner, F., Lair, G.J. and Gerzabek, M.H. (2009a) Rapid carbon accretion and organic matter pool stabilization in riverine floodplain soils. Global Biogeochem. Cycles, **23**, GB4004, doi:10.1029/2009GB003481.

Zehetner, F., Lair, G.J., Graf, M. and Gerzabek, M.H. (2009b) Rates of biogeochemical phosphorus and copper redistribution in young floodplain soils. *Biogeosciences* **6**, 2949–2956.

Chapter 7

BASIN – Integration of research in AquaTerra river basins

J. Joziasse, N. Baran, C. Mouvet, M. Petrovic, D. Barceló, J. P. M. Vink, S. Brouyère, Ph. Orban, J. Batlle-Aguilar, I. Lobe, W. von Tümpling, F.-J. Maringer, J. Slobodnik and H. H. M. Rijnaarts

KEY OBJECTIVES

The objective of sub-project BASIN is to connect the scientific research in the sub-projects HYDRO, BIOGEOCHEM, FLUX, COMPUTE, and TREND to practical research cases (information from DOW). BASIN is divided into five work packages, related to five European river basins differing in climate, size, functioning, and pollution profile. The objectives of the work packages are listed in Table 7.1.

KEY ACHIEVEMENTS

Results brévilles

The local spring, previously an important source for local water supply, was disconnected from the distribution network in August 2001 because pesticide and nitrate concentrations exceeded drinking water quality limits. Continued research on this system showed that pesticides such as atrazine (At) with an application stopped in 1999 (four years before the official ban in France, thanks to the collaboration of farmers located through the catchment) can be found with its metabolites and especially deethylatrazine (DEA) in groundwater almost ten years after the last application of At and in the spring without any significant decrease for DEA (Figure 7.1). The data (raw data and estimated outfluxes – see Figure 7.1 and Figure 7.2) discussed in detail in deliverable F1.16 (Baran *et al.*, 2007) suggest persistence and slow transport of this molecule and its related compounds in the subsurface (Barth *et al.*, 2008). The fluxes of (At + DEA), representing less than 500 g/year lead to concentrations higher than the drinking water level.

Results Ebro

A GIS system has been developed for interpretation of historical analyses and monitoring data generated in a three year surveillance monitoring programme (Terrado *et al.*, 2006). In surface water, the most important pollution problem was caused by a contamination pattern from alkylphenols (APs), substances originated by industry and urbanisation.

 Advanced Tools and Models to Improve River Basin Management in Europe in the Context of Climate Change (AquaTerra) by Michael Finkel, Johannes Barth and Peter Grathwohl ISBN: 9781843393726. Published by IWA Publishing, London, UK.

In sediments, two main contamination patterns, one loaded by hydrocarbons (PAHs) and the other by the group of organochlorinated compounds (OCs), were identified as the main contamination problems. Wastewater treatment plants (WWTP) were identified as sources for aquatic contamination by pharmaceuticals in the Ebro River basin, see Figure 7.3. (Gros *et al.*, 2007, Gros *et al.*, 2010). Conventional wastewater treatment applied in seven WWTP in the Ebro river basin is found to be unable to efficiently remove a wide range of pharmaceuticals studied. Hazard posed by pharmaceuticals in both surface and effluent wastewaters was assessed toward different aquatic organisms, (algae, daphnids and fish). Results indicate that no significant risks could be associated to the presence of pharmaceuticals in those matrices, indicating that reduction of compound concentration after wastewater treatment as well as the dilution once pharmaceuticals are discharged in receiving river water efficiently mitigate possible environmental hazards.

Table 7.1 The objectives of the BASIN work packages.

Work package	Main objectives
Brévilles (R1)	To integrate and quantify perturbations from land use (agriculture) on groundwater quality resulting from a major change in agricultural practices (substitution in year 2000 of atrazine by a recently registered pesticide: acetochlor).
Ebro (R2)	Analysis of vulnerability of the Ebro basin according to the existence of industrial activities and cultivated land and the following parameters: for groundwater, main interest is focused on water table aquifers (porous and karstic), for surface water circulating flow in the river, runoff coefficient and importance of soil erosion.
Meuse (R3)	For the Dutch Meuse basin the development of a spatial, interactive decision support system (DSS) to calculate the chemical speciation of heavy metals in floodplain sediments under various environmental conditions. Effects of redevelopment measures by scenario comparisons. Furthermore, to improve knowledge about soil, sediment, groundwater and river quality in the tributary Dommel. For the Belgian part of the Meuse system quantification of groundwater quality effects by river – groundwater interaction and quantification of coupled ecotoxicological effects of contaminants from sediment, suspended solids and freshwater on aquatic organisms.
Elbe (R4)	The integrative investigation and determination of the fate and transport of typical organic pollutants such as PCB's, HCH's and DDT, as well as their main metabolites and inorganic contaminants such as Hg, As and Pb. The investigation includes suspended matter transport via sediments and subsurface soils up to effects for the groundwater in the Elbe river basin during the landscape renovation and changes of land use at selected sampling sites. Verification of biodegradation processes of persistent organic contaminants, in close co-operation with BIOGEOCHEM.
Danube (R5)	The relation between land-use and suspended solid loads in the river system, with special attention to river segments in Austria and Hungary. The fate and transport of important chemicals, and the filter buffer, transform, function of the soil-sediment-river system in the investigated Danube transect are related to spatial and temporal changes in land-use practises. In addition, investigation of ecotoxicological impacts of contaminants, associated with sediments and suspended solids, using biological test- and monitoring methods and establishment of a sediment-, water- and plant-test-system on a research vessel, for the investigation of toxicity and bioavailability of particle-bound toxicants.

In the Cinca area, contamination sources of brominated flame retardants (BFRs) were identified (Eljarrat *et al.*, 2007).

Results Meuse

Application of the BioChem model: case study Dommel basin

The BioChem-decision support system (DSS) was specifically designed for the prediction of heavy metal availability in floodplain sediments under varying environmental conditions (e.g., redox potential). For example, the model can be used for prediction of changes in metal availability as a result of various inundation or sedimentation scenarios. It consists of a chemical module and an ecotoxicological module. The chemical module quantifies the pore water concentration of heavy metals depending on a number of physical and chemical parameters determining the speciation of the various system components. The ecotoxicological module quantifies the biological availability of metals in selected organisms (plants and macrofauna organisms). A case study with two sedimentation scenarios was carried out in a flood plain site of the Meuse tributary Dommel, see Figure 7.4. The actual risks of heavy metal availability at the site depend on the type of land use and habitat, and key species that may occur or are foreseen in future redevelopment situations.

Medio 2008 the input manager was renewed (project deliverable R3.28, see www.eu-aquaterra.de). This significantly improved the stability of the calculation module. The new manager has a more advanced modularity; procedures, such as the schematization of soil layers and the selection of chemical compounds, are better structured and less vulnerable to user-related errors.

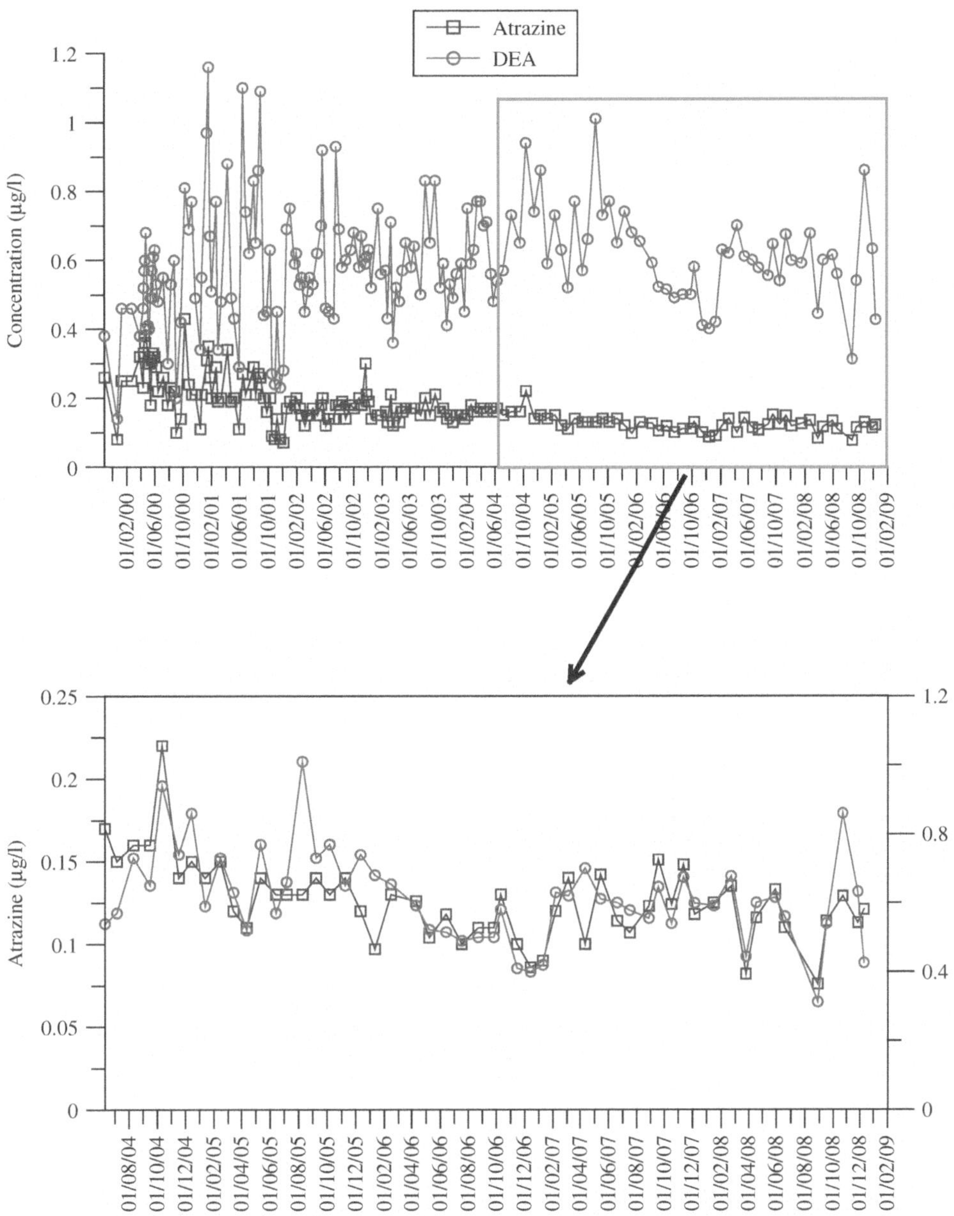

Figure 7.1 Atrazine and deethylatrazine (DEA) concentrations measured at the Brévilles spring since October 1999 and during the AquaTerra project.

A new water compartment module was built, enabling calculations of water chemistry based on suspended matter interaction rather than via the schematization of sediment. This module will be extended with a state-of-the-art risk assessment concept, i.e., the biotic ligand models for metals (BLM). BLMs are capable of normalizing generic quality standards, based on the composition of surface waters. BLMs for various metals are optimized and operationalized to operate within this module of BioChem.

Ecotoxicological research in the Walloon Meuse river basin

Bioassays and chemical analyses were used to quantify coupled ecotoxicological effects of contaminants from sediment, suspended solids and freshwater on aquatic organisms. For this ecotoxicological research, five sampling points were selected within the Walloon part of the Meuse basin, including two tributaries, the Sambre and the Vesdre. Sediments and suspended solids were collected and analysed using a battery of bioassays that represent different trophic levels and different functions, exposure routes and sensitivities. Complementary chemical analyses were performed in order to study the degree of metal contamination and to support the interpretation of endpoints results. Biogenic capacity can be considered as an expression of toxicity and the nutritional value of suspended solids. The investigations proved that the gradient of pollution from upstream to downstream in the Meuse system is reflected in the ecotoxicological effects studied.

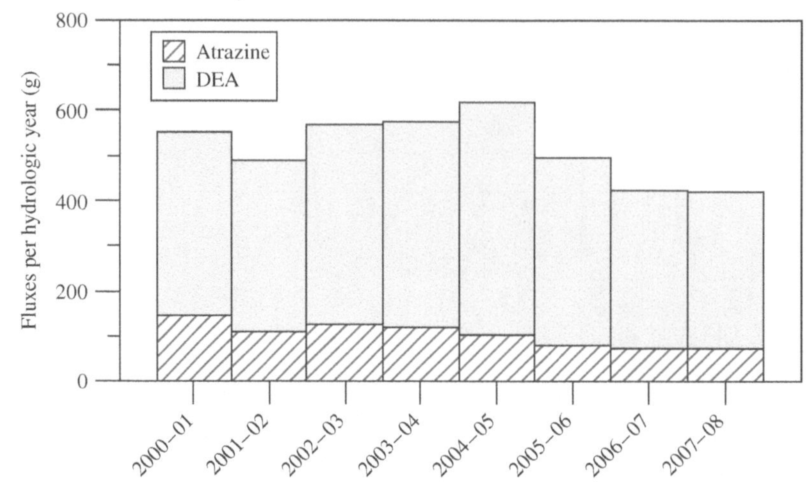

Figure 7.2 Atrazine and deethylatrazine (DEA) fluxes per hydrological year exported at the Brévilles spring.

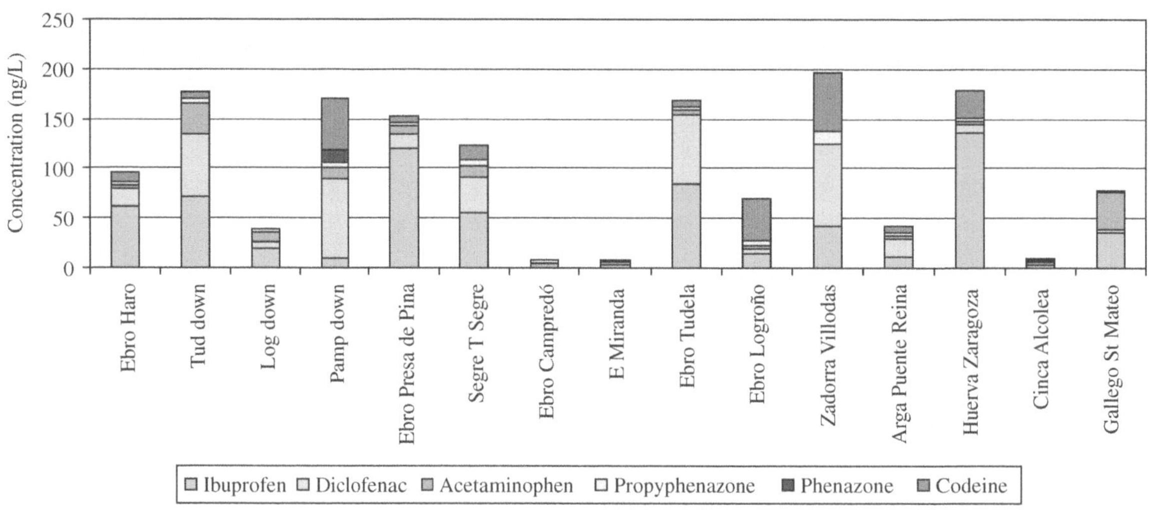

Figure 7.3 Concentrations of the most ubiquitous analgesics/anti-inflammatories detected in river water downstream of the wastewater treatment plants (WWTP) monitored along the Ebro river basin in Spain.

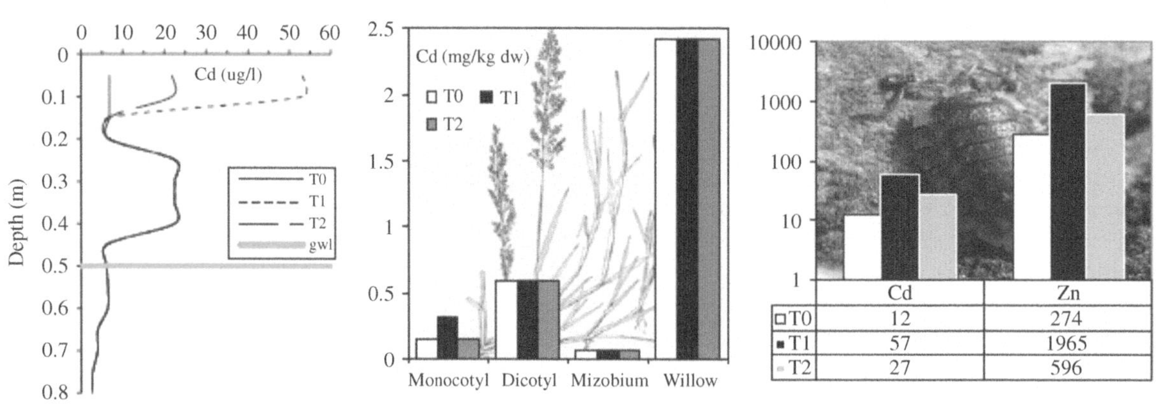

Figure 7.4 Some results of the Dommel case study. The T1 (10 cm sedimentation of the poor quality Dommel sediment) results in an increase of all metal concentrations, except arsenic, in the topsoil compared to the present (T0) situation. The T2 scenario (metal load from the Eindegatsloop is reduced to zero) leads to less excessive concentrations in the pore water of the topsoil. For most plant groups, a sedimentation layer of 10 cm is too small to achieve physical contact between sediment and roots. For monocotyls (grasses) however, some differences occur. Grasses are a preferred food category for most herbivores. Both scenarios for sedimentation lead to an increase of body concentrations of invertabrates, as is shown here for Porcelio scaber. This leads to a potential increase of metals in the food chain.

Dispersion of contaminants from groundwater into surface water: the Flémalle test site

A methodology for the quantification of contaminated groundwater discharge in surface water (groundwater flow and transport model) has been developed at the Flémalle test site. This is an industrial brownfield area along the Meuse of 7.3 ha, located on the left bank of the river Meuse, upstream from the city of Liège. Past industrial activities have produced an important contamination of soil, subsoil and groundwater.

Former field investigation campaigns were performed from 1984 to 2005, based on sampling and drilling of boreholes and piezometers. These investigations have allowed to delineate and quantify the nature, extent and level of contaminations for a large variety of contaminants, such as BTEX, PAH, cyanides, heavy metals, mineral oils, in both the unsaturated zone and the saturated alluvial deposits.

In the scope of AquaTerra, further investigations have been performed in the brownfield, focussing more on the alluvial aquifer. These investigations consisted of regular piezometric surveys, continuous monitoring of surface water level in the Meuse river and groundwater levels at various piezometers, at high frequency and over relatively long periods of time, pumping tests, infiltration tests and single/multiple well tracer tests (Batlle-Aguilar, 2008). Data on rainfall have also been obtained. More specific investigations were performed by other AquaTerra partners, to study the specific behaviour of contaminants such as mono-aromatic and polyaromatic hydrocarbons (microcosm studies and isotopic fractionation by CHYN), heavy metals (lab investigations by VITO).

Field investigations and measurements provided a better knowledge on the hydrogeological conditions, on groundwater-surface water interactions and on the specific behaviour of contaminants existing in the brownfield (sorption, degradation as a function of biogeochemical conditions prevailing in the field).

In addition, a detailed groundwater flow and transport model was developed, using Modflow2000 and MT3DMS. This model was created using an original two-scale combined zonation (regional scale) – pilot point (local scale) approach. Groundwater flow conditions were calibrated under transient conditions, using the detailed dataset obtained with groundwater – surface water monitoring performed in the field. Among others, this provided a very detailed heterogeneous hydraulic conductivity field for the alluvial deposits corresponding to the brownfield (Batlle-Aguilar, 2008). The transport model was calibrated using tracer test results, performed in the field. Various contaminant dispersion scenarios were modeled for benzene, which was identified from specific lab and field investigations as the potentially most problematic contaminant present in the alluvial aquifer, in terms of risk of dispersion off-site, in the direction of the Meuse river. These scenarios indicate that biodegradation under sulfate-reducing conditions is the key process that seems to prevent benzene to migrate off-site (Batlle-Aguilar, 2008).

The AquaTerra activities concerning the Flémalle site in 2008 have been devoted to delivery and valorization of the practical results by end-users: the Walloon Environmental Institute (SPAQuE).

A new hydrological model to deal with climate change impacts assessment: the Geer basin

Specifically for the Geer basin (tributary of the Meuse river), a new hydrological model has been developed to deal with climate change impacts assessment. The model fully integrates surface and subsurface water, especially focussing on recharge processes, which are key elements in the context of groundwater management and changing climate. The work in 2008 focussed on the further improvement of the Geer basin model, with specific attention to the calibration. The impacts on the spatial and temporal discretisation on the model results have been tested and also climate change scenarios, simulated by Newcastle University, have been extensively tested on the Geer basin model. This work has been done in cooperation with the HYDRO subproject, as support to the the Groundwater Division of the Administration for Agriculture, Environment and Natural Resources of the Walloon region (DGARNE) for the implementation of the Water Framework Directive (WFD).

At present, nitrate concentrations in the Geer basin are close to the drinking water limit (50 mg/l). Estimation of present and future groundwater quality trends in this basin is of the highest importance for supporting any decision in terms of land use (changes in agricultural practices, etc.). Specific investigations consisted of a regional groundwater sampling campaign for measuring nitrate and tritium contents in groundwater. This campaign was also partly financed by DGARNE. These measurements have allowed a better understanding of the hydrogeological context of the Geer basin and of the regional distribution of nitrates which are mostly related to groundwater age and mixing between old and young groundwater. Three zones have been identified. The first zone, located in the south-western part of the basin, corresponds to the groundwater recharge area, with the highest nitrate contents observed in the basin. The second zone, located in the north-eastern part of the basin, is the groundwater discharge area (mostly in the Geer river), with a mixing between old and young groundwater and intermediate nitrate contents. The third zone, corresponding to the extension of the chalk aquifer to the north-west, outside the Geer basin, is characterized by old groundwater, almost uncontaminated with nitrate.

All available data were used to develop a variably saturated groundwater flow and solute transport model using the Hybrid Mixing Cell Finite Element method as coded in the HGULg finite element code SUFT3D. This model was calibrated in steady state groundwater flow conditions using two contrasted high and low piezometric situations. The regional solute transport model was then calibrated using the tritium dataset. The calibrated model was used to assess (first) and to forecast (next) nitrate trends in groundwater of the chalk aquifer, based on a uniformed and simplified, but yet realistic nitrate input function in the groundwater recharge.

The modelling results show that the transport model was able to very well reproduce observed trends and present nitrate concentration levels observed in the different parts of the Geer catchment. The model was also used as a support for a socio-economic analysis of the problem generated by the presence of nitrate in groundwater of the Geer basin. This analysis consisted of a cost-efficiency and cost-benefits of programmes of measures defined to reverse nitrate trends and to result in nitrate concentrations less than 50 mg/l in the future.

Results Elbe

At the Schachtgraben site, research results indicate the release of dissolved contaminants from the streambed sediments, which poses a risk of the downstream propagation of contaminants (Figure 7.5). Besides, the annual down cutting of the riparian vegetation significantly mobilizes streambed sediments. In the Spittelwasser, polluted sediments currently deposited in the floodplains, cover old, more polluted sediment deposits. However, as the maximum concentrations were found in the 5–15 cm segments of the topsoil, every construction activity, or movement of the topsoil layer in the floodplains or the river bank can lead to a remobilization of polluted sediments which are currently protected against erosion through the root network. The sediments deposited in the floodplain pose a potential risk for the groundwater at zones of higher pollutant concentrations. At the Schoenberg site, soil solution samples revealed a substantial mobilization of β-HCH and HCB in the course of a flood event. The parallel increase of dissolved organic carbon (DOC) suggested a co-release with DOC. Cd, Cu and Zn were immobilized due to the inundation. The release of Co and Mo was assumed to be the result of the dissolution of Mn and Fe oxides and hydroxides at lower redox potentials. The release of Ni and Cr was associated with DOC. Results indicated that a significant release of As, Cr, Ni and DOC needs an inundation of the floodplain longer than six weeks.

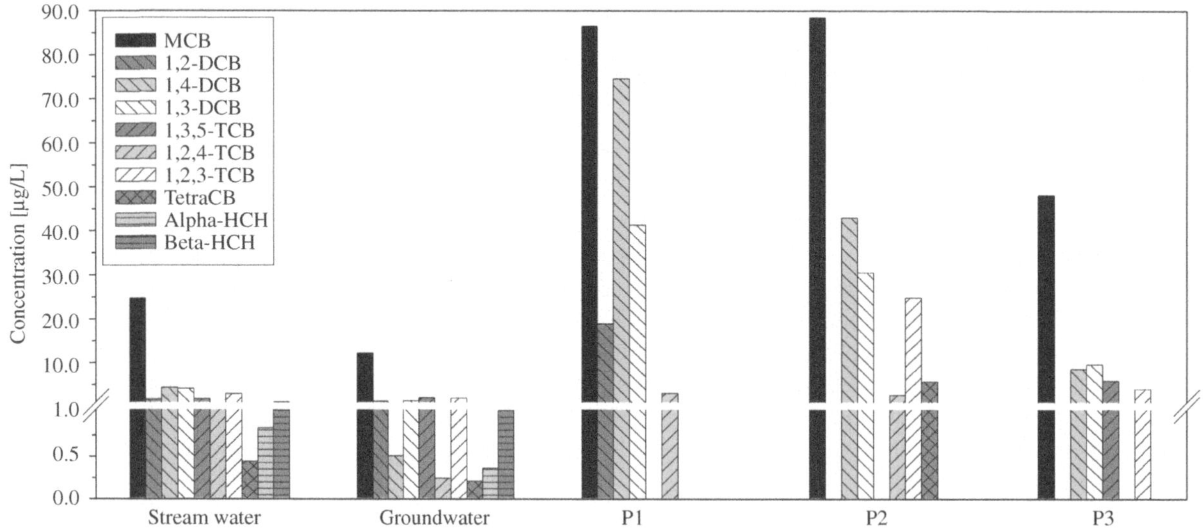

Figure 7.5 Comparison of the contaminant concentrations at the Schachtgraben site in the pore water of the streambed at the sampling points P1 (medium groundwater discharge), P2 (low groundwater discharge) and P3 (high groundwater discharge) with the average concentrations in the surface water and in the groundwater. The streambed sediments are more heavily contaminated than the groundwater and the surface water. The large amounts of contaminants may have been transported to the streambed via diffusion and hyporheic exchange and may have adsorbed to the sediments over a long time period. Nowadays, the contaminants may desorb into the discharging, less contaminated groundwater and may be transported into the stream by this mechanism.

Results Danube

The AquaTerra Danube Survey (ADS) has resulted in the collection of a large number of sediment, suspended matter and water samples from a 1147 km long Danube stretch starting in Klosterneuburg (Austria, river km 1942) and ending in Calafat (Romania, river km 795). The analyses results were organised in a web-based database and compared with earlier data. Analyses of the data in the ADS-database led to the identification of relevant pollutants in suspended matters, sediments and water.

Radio-isotopic characterisation of sediment and suspended particles provided knowledge about the age and origin of the samples. The main part of the contamination by artificial radionuclides (e.g., ^{137}Cs) results from Chernobyl fallout and subsequent soil erosion of solid particles into the runoff. The relatively high ^{137}Cs Chernobyl contamination of soil in the upper part of the Danube catchment has been dispersed downstream into the bottom sediments of the lower part of the river. There is a medium term (10–20 years) influence of upstream pollutants sources on sediment burden downstream. Metals/elements of more geochemical than anthropogenic origin show similar ecological behaviour in solid particles as natural occurring radionuclides. The results for bottom sediments give an indication of an anthropogenic soil surface erosion part of some metals that increases in the flow direction. A main characteristic of various organic pollutants and ^{137}Cs is the high affinity/binding capacity with soil clay particles. Due to this fact, positive (linked) correlations between analyzed organic molecules and ^{137}Cs have been observed in this investigation.

This behaviour means similar ecological erosion and transport processes of anthropogenic organic pollutants and ^{137}Cs contamination of soil particles. Generally, negative correlations were found between metals and ^{137}Cs (see Figure 7.6). These correlations are strongly influenced by the negative correlation of ^{137}Cs with sampling location in the flow direction. The those two correlations together indicate an increasing anthropogenic/soil surface erosion part of metals in the flow direction.

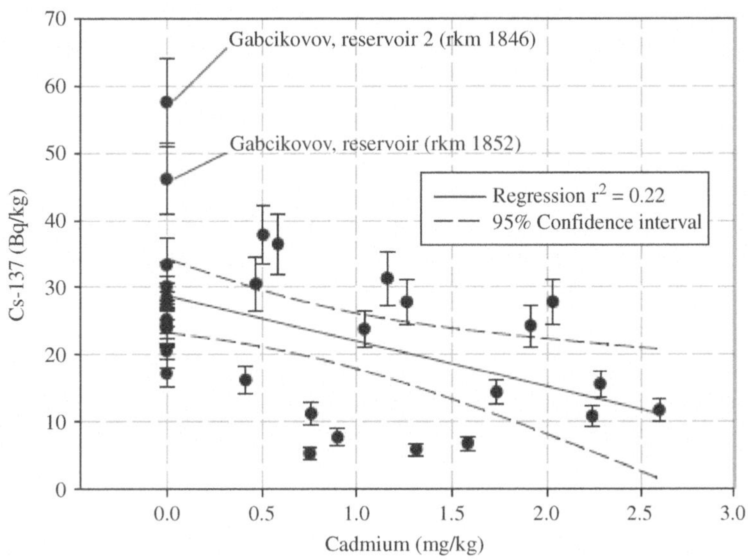

Figure 7.6 Cadmium belongs to the metals for which the most significant negative correlations with ^{137}Cs activity occur.

The Danube survey had a significant positive impact on developing contacts with stakeholders and promoted collaboration of the International Commission for the Protection of the Danube River (ICPDR) in other projects. The survey raised scientific awareness of stakeholders and raised stakeholder interest for the results of the Danube survey. The outcome of a Danube workshop in 2006 is that the participants, originating from all 13 major Danube countries, committed themselves to draft and support a roadmap towards an advice on the implementation of sediment management in the Danube Water Framework Directive (WFD) river basin management plan.

IMPLICATIONS AND APPLICATION OF RESULTS

Locally

The results of the Brévilles site suggest persistence (during several years) and slow transport of atrazine and its main metabolite DEA. Besides, it takes time before before measures to protect water quality show effect. Local water managers and policy makers should take these facts into account in their plans for water use in the future. The ten years monitoring period reveals that the hydrological cycle is not completed during this period, suggesting that the monitoring should necessarily be longer than this period to fully understand the fate of pesticides and predict trends for groundwater quality.

Ebro data of AquaTerra for priority substances can be compared with the data obtained by CHE (Confederacion hidrografica del Ebro, river basin management authority), since the same sampling points were used, allowing an inventory of contaminants and monitoring of levels and trends. Analysis of industrial effluents revealed that the industry which produced EPS (expandable polystyrene) treated with flame retardants and ABS (acrylonitrile-butadiene-styrene) is the likely responsible cause of the hexabromocyclododecane (HBCD) contamination of the Cinca river, a tributary of the Ebro River. The outcome of the research was used to fix emission limits for industry (PVC, textile in the Cinca area).

The outcomes of model calculations with BioChem and other instruments such as applied in the Meuse basin, raise important information on the availability of contaminants in pore water (and thus the risks involved) under various scenarios of inundation frequencies, contamination reduction measures, etc. In case the heavy metals emissions of the industrial source involved shall be stopped, this will eventually lead to a significant reduction of the ecological risks associated with deposition of contaminated suspended matter in floodplain soils. The knowledge obtained from scenario calculations can be used by river basin managers to estimate the effect of several types of measures, such as redevelopment projects and selection of water retention areas. In the groundwater flow and transport model for the Flémalle cokery site (including the interaction with the Meuse), transport and attenuation of contaminants (mostly organics – benzene and PAH) in the gravel aquifer is described. This application can be used by the Wallonian environmental institute SPAQuE, for decision support in terms of groundwater remediation. The hydrological model developed in the Geer basin can be used for estimation of the impacts of groundwater extractions and for the evaluation of the nitrate contamination in the regional groundwater.

The knowledge of contaminant behaviour of typical pollutants in the Elbe catchment is valuable for river basin managers, when developing measurements and regulations. AquaTerra research has resulted in some recommendations for stakeholders in the region (see further below).

The AquaTerra data obtained for Danube can be added to the JDS monitoring data. The complete dataset can be used for trend calculation, that will provide insight in effects of measures for the improvement of the water quality in the Danube basin.

In a European and a global context

The Brévilles research provided knowledge about the behavior of pesticides from soil to and into groundwater. Scaling up the pollution problems to catchment and basin scales would likely cause even longer time periods for pesticide turnover in subsurface environments (Barth *et al.*, 2008). Integrated monitoring of groundwater quality coupled with the use of different tools (modelling, isotope geochemistry, tracer tests, geophysics, etc.) is a very promising scientific approach (Gutierrez and Baran, 2009). The implementation of the WFD will also require this type of approach on a detailed study scale (Baran *et al.*, 2007) in order to appreciate the complexity of monitoring and representativeness of measurements (date of measurements, point of observation). The persistence of atrazine several years after the last application and the fact that the new used molecule was not detected to date, is particularly interesting in term of management of the aquifers and within the framework of the evolution of the regulation.

In the Ebro work package the coupling of chemometric data analysis results with the use of geostatistical methods is shown to be a powerful procedure to underline the more significant spatial and monthly variations of the contamination patterns, taking into account the particular geographical structure of the area under study. Patterns of contamination by pharmaceutically active compounds are applicable to other river basins. In the future, environmental data should be compared to regional data on consumption and use.

Besides for the Meuse, the BioChem model can be used to identify ecological risks posed by heavy metals in floodplain soils of other river basins as well. Furthermore, it sustains the development of standards that are more based on (ecological) risks than on absolute contaminant contents in (floodplain) soils. Bioassays, as evaluated in the Meuse work package, may be considered as a validation tool or a complement to simple partitioning models. The monitoring sampling scheme as applied in Belgium and the Netherlands is also applicable in other river basins. The same is true for the application of models that integrate groundwater and surface water contaminants transport.

The comparison of contaminants in pore water of the streambed, in surface water and groundwater (as has been applied for the Elbe catchment) provides insight into the risks of contaminants mobilisation, especially when large-scale redevelopment projects are carried out.

The river characterisation methods applied in the Danube work package, such as isotopic and chemical characterisation of sediment and suspended particles can provide insight into erosion and transport processes of pollutants. This can be used for the EU WFD analysis of river basin characteristics.

CONTRIBUTIONS AND POLICY RECOMMENDATIONS FOR RIVER BASIN MANAGEMENT

Locally/nationally

The national ban of atrazine takes place four years after the last application in the Brévilles site. The demonstration of the persistence of contamination ten years after the stop is of interest for the water management. Indeed, the improvement of groundwater quality could be notably delayed to the changes of practices. In terms of survey (Water Framework Directive, drinking water, etc.), these aspects have to be considered.

In the Ebro river basin, analysis of industrial effluents revealed significant hexabromocyclododecane (HBCD) contamination of the Cinca river, a tributary of the Ebro river, most likely originating from the industry which produced EPS (Expandable polystyrene) treated with flame retardants and ABS (Acrylonitrile-butadiene-styrene). The outcomes of the research were used by the river basin managemers (Confederacion Hidrografica del Ebro, CHE) to fix emission limits for industry (PVC, textile in the Cinca area)

BioChem DSS results for the Dommel catchment showed that heavy metal availability is related to ecological risk (if the heavy metals are present in the pore water, they can be taken up in the food chain). Therefore, it is necessary to take into account and quantify the availability of heavy metals (and other contaminants) in floodplain soils when planning and managing land-use.

For the end-users of the results at the Flemalle-site, a compilation of specific final reports has been prepared in French, as a support to decision-making. SPAQuE uses the results for the definition of a rehabilitation programme for the Flémalle site. Based on the AquaTerra results and associated recommendations, the SPAQuE society is now evaluating several alternatives for groundwater cleanup, including monitored natural attenuation after removal of the organic contaminant sources.

All results and conclusions from the Geer study site have been delivered to the Walloon region authorities and to the main water company active in the Geer basin in the form of a detailed summary report written in French. The Walloon region authorities have already used trend analysis results for their reporting to the EU as arguments for obtaining derogations for the 2015 WFD demands. Recent modelling results are now taken into account by the Walloon region and by the Liège water company Compagnie Intercommunale Liégeoise des Eaux (CILE) as a support to decision making for the management of the chalk aquifer (management plans, necessity for groundwater cleanup scenarios, etc.).

In a European context

As a result of the Water Framework directive, atrazine is banned in many European countries. For example, its use in France was banned in 2003. From a European regulatory perspective, the Brévilles site provides a valuable case study on the improvement of groundwater quality following a change in agricultural practices. The key findings have been that contamination of groundwater in the Brévilles aquifer by atrazine and its metabolite, deethylatrazine, persists for several years after the last use of atrazine and that their presence will probably continue for several years to come. This is a result of the intrinsic properties of the molecules and the hydrodynamics of the site, such as the buffering effect of the soil, the unsaturated zone and the long transfer time through the aquifer. A further significant finding is that neither the substitute molecule, acetochlor, nor its metabolites has been detected in groundwater as yet. Several theories are available to explain this fact, but further monitoring and more time will help to confirm which of these are most likely. Consequently, it is recommended to investigate transfer times of pesticides in groundwater in order to provide more insight in the time it will take for contaminated aquifers to return to the desired water quality.

The development of the methodology dealing with quantification of the flux from groundwater to surface water (FLUX, TREND and BASIN SPs in the Meuse basin), relates to the fact that fluxes are highly dynamic. Therefore, it is recommended to undertake a monitoring of the groundwater and surface water levels on hourly basis and for a duration of approximately a month (at each monitored location in the aquifer) in order to determine the flux between surface water and groundwater. This monitoring is not necessarily expensive, as it can be done with water level probes coupled with an automatic data log interface and passive samplers as continous monitoring devices. The monitoring can be coupled with tracer testing in order to improve accuracy of flux quantification.

BioChem results show river basin managers and policy makers the factors that influence risks of heavy metals, which is important for planning and managing land-use. When selecting water retention areas, it is recommended to consider suspended matter quality and the effect of flooding on metal availability and the resulting potential bioaccumulation. Therefore, in monitoring campaigns more attention should be paid to the monitoring of suspended sediment quality.

Regarding the prevention of contaminant migration, river basin managers should consider land-use with vegetation schemes in order to avoid erosion. Within the framework of the Elbe research, it is proposed to the stakeholders that a less radical cutting of vegetation method would prevent the mobilization of contaminated sediments. Some pollutants partly exceed the control value for the path soil-groundwater for a limited period during inundation of the floodplain or shortly afterwards. It is recommended that the filter and retention function of the soil should be further controlled. It is equally necessary to continuously investigate sediment quality in addition to water quality in order to assess soil and groundwater quality and evolution. Quantification of fluxes is very important, despite the challenge that fluxes become increasingly difficult to measure during flood events. These activities contribute to the characterisation of the river system. It is recommended that phyto-remediation could be used to reduce contaminant migration during floods events (through metals mobilisation in plants and reduction of water infiltration). The vegetation could then be used for agriculture or as carbon source for biomass.

In order to understand natural resilience of the system towards contamination, it is furthermore necessary to establish long-term environmental monitoring. This enables river basin managers, policy makers and researchers to understand contamination behaviour in the sediment. The use of isotopic analysis is recommended because it can distinguish anthropogenic and natural sources of contamination, which makes it possible to correlate sediment quality and surrounding land-uses. Continuation of cooperation with river basin authorities and other stakeholders in the field of monitoring and river basin management is encouraged. One of the issues that need continuous attention is harmonisation of sampling and analysis methods.

OPEN & NEW QUESTIONS FROM BASIN

For the Brévilles spring, it was impossible to predict how long it will take to return to a water quality that meets drinking water standards, the delay depending on intrinsic properties of molecules but also on the recharge linked to climatic conditions. Mechanisms of water and solutes infiltration, solute remobilization and the aquifer system must be studied in more detail. From a hydrodynamic point of view, we must improve our knowledge of transfer times in the vadose zone and the saturated zone. More studies specific of pesticides must be done in areas such as desorption kinetics coupled with the knowledge of stocks of At and DEA in the soil and in the vadose zone. All of these developments would enable us to model the transport of pesticides at the scale of an aquifer such as the one studied here (Morvan *et al.*, 2006).

For risk assessment and ecotoxicological implications of pharmaceuticals as environmental contaminants, in-depth evaluation of risks is needed. Unless more is known about the possible chronic effects of individual pharmaceuticals and mixtures, conclusions on the hazards or risk of pharmaceuticals (e.g., in the Ebro Basin) to the aquatic ecosystem are premature yet.

More knowledge is necessary on confounding factors for the bioassay test systems. Confounding factors (e.g., dissolved oxygen, pH, ammonia and conductivity) are factors that interfere within the toxicity results and mask the real cause of the toxicity. Suspended solid research needs to be further developed, as we assume a significant transport in this phase, particularly under flooded conditions. Special attention needs to be paid to the relation between sampling, preservation and toxicity. Besides, cost effectiveness of bioassays in monitoring programs more such integral tests need to be applied in the field and set into relation to contaminant matrices.

An important question for the Elbe river basin (but also for other river basins such as the Meuse) is the quantification of contaminants transport in the basin, especially sediment-bound contaminants. This calls for development of 2D/3D modelling tools that describe sedimentation and resuspension of sediment particles in the streambed and in the floodplains under various discharge situations.

REFERENCES

Baran, N., Mouvet, C. and Négrel, Ph. (2007) Hydrodynamic and geochemical constraints on pesticide concentrations in the groundwater of an agricultural catchment (Brévilles, France). *Environmental Pollution* **148**, 729–738.

Barth, J.A.C., Grathwohl, P., Fowler, H.J., Bellin, A., Gerzabek, M.H., Lair, G.J., Barceló, D., Petrovic, M., Navarro, A., Négrel, Ph., Petelet-giraud, E., Darmendrail, D., Rijnaarts, H., Langenhoff, A., de Weert, J., Slob, A., Frank, E., Gutierrez, A., Kretzschmar, R., Gocht, T., Steidle, D., Garrido, F., Jones, K.C., Meijer, S., Moeckel, C., Marsman, A., Klaver, G., Vogel, T., Bürger, C., Kolditz, O., Broers, H.P., Baran, N., Joziasse, J., von Tümpling, W., van Gaans, P., Merly, C., Chapman, A., Brouyere, S., Batlle, Aguilar, J. and Orban, Ph. (2008) Mobility, turnover and storage of pollutants in soils, sediments and waters: achievements and results by the EU project AquaTerra. A review. Invited publication in Agronomy for Sustainable Development, Agron. Sustain. Dev. 28 (2008) INRA, EDP Sciences, DOI: 10.1051/agro:2007060.

Batlle-Aguilar, J. (2008) Groundwater flow and contaminant transport in an alluvial aquifer: in-situ investigation and modelling of a brownfield with strong groundwater – surface water interactions. Ph.D Thesis. Applied Sciences Faculty, University of Liège. Liège (Belgium). pp. 245.

Eljarrat, E., Labandeira, A., Marsh, G., Raldúa D. and Barceló. D. (2007) Decabrominated diphenyl ether in river fish and sediment samples collected downstream an industrial park. *Chemosphere* **69**, 1278–1286.

Gros, M., Petrovic, M. and Barcelo, D. (2007) Wastewater Treatment Plants As A Pathway For Aquatic Contamination By Pharmaceuticals In The Ebro River Basin (Northeast Spain). *Environ. Toxicol. Chem.*, **26**(8), 1553–1562.

Gros, M., Petrović, M., Ginebreda, A. and Barceló, D. (2010) Removal of pharmaceuticals during wastewater treatment and environmental risk assessment using hazard indexes. *Environment International* **36**(2010), 15–26.

Gutierrez A. and Baran N. (2009) Long-term transfer of diffuse pollution at catchment scale: respective roles of soil, and the unsaturated and saturated zones (Brévilles, France). *Journal of Hydrology* **369**, 381–391. doi:10.1016/j.jhydrol.2009.02905.

Morvan, X., Mouvet, C., Baran, N. and Gutierrez, A. (2006) Pesticides in the groundwater of a spring draining a sandy aquifer: temporal variability of concentrations and fluxes. *Journal of Contaminant Hydrology* **87**, 176–190.

Terrado, M., Barcelo, D. and Tauler, R. (2006) Identification and distribution of contamination sources in the Ebro river basin by chemometrics modelling coupled to geographical information systems. *Talanta* **70**, 691–704.

Chapter 8

COMPUTE – Modelling methods and strategies

A. Bellin, B. Majone, M. Zambrano-Bigiarini, E. Bertuzzo, G. Botter, C. Bürger, A. Gutierrez, O. Kolditz, M. Marani, A. Perez-Garcia, A. Rinaldo and T. Vogel

KEY OBJECTIVES OF COMPUTE

The major objective of COMPUTE is to enhance our understanding of soil-water ecosystems at the catchment scale. In particular, system responses to changes in climatic and man-made conditions (e.g. land use) have been studied in order to assess the impacts on water quantity and quality at different investigation scales ranging from hillslope to catchment and river basin scales. For this purpose appropriate up-scaling methodologies have been developed in order to bridge across this multiplicity of scales (see Figure 8.1).

Specific objectives were the following:

- to quantify the impact of actual and future climatic conditions on quality and availability of water resources;
- to quantify the impact of changes in land use and soil functioning on quality and availability of water resources;
- to identify significant changes and trends in the water cycle at the basin scale, both for surface and subsurface water resources.

KEY ACHIEVEMENTS

We developed new modelling strategies and approaches to investigate hydrological processes at the three main scales of interest, i.e. the hill slope, catchment and river basin scales. At the hill slope scale (COMPUTE 1) we investigate local processes and the role of heterogeneity of soil properties in controlling runoff production and solute transport. The main focus is on vertical soil water dynamics including preferential flow effects. The spatial scales of interest range from meters to hundreds of meters. In the workpackage COMPUTE 2 the same processes are upscaled to the catchment in order to describe the relevant processes that cooperate to generate runoff and transport contaminants over scales ranging from hundreds of meters to tens of kilometres, in a context with similar climate and geomorphologic characteristics. The workpackage COMPUTE 3 considers a third level of analysis involving much larger spatial scales, say catchments which size exceed 1000 km^2, where hill slope dynamics combine with processes occurring within the river network and where anthropogenic impacts, in terms of water withdrawals, transfers of waters from a catchment to another, and release of contaminants of both distributed and concentrated origin, assume a critical role in controlling the state of the system. The key element that separates river basin from catchment scale modelling is the variety of climate and geomorphological properties that characterize the latter with respect to the relative homogeneity of the catchment scale. In essence, the larger spatial scale of the river basin approach is the direct consequence of this variety, rather than a strict rule based on a threshold on the spatial extension of the domain. Finally, workpackage COMPUTE 4 provides a software framework to

 Advanced Tools and Models to Improve River Basin Management in Europe in the Context of Climate Change (AquaTerra) by Michael Finkel, Johannes Barth and Peter Grathwohl ISBN: 9781843393726. Published by IWA Publishing, London, UK.

combine the variety of modelling tools developed within the WPs and sets up a common data base for an efficient exchange of data and information. In the ensuing section we provide a detailed description of COMPUTE's main achievements.

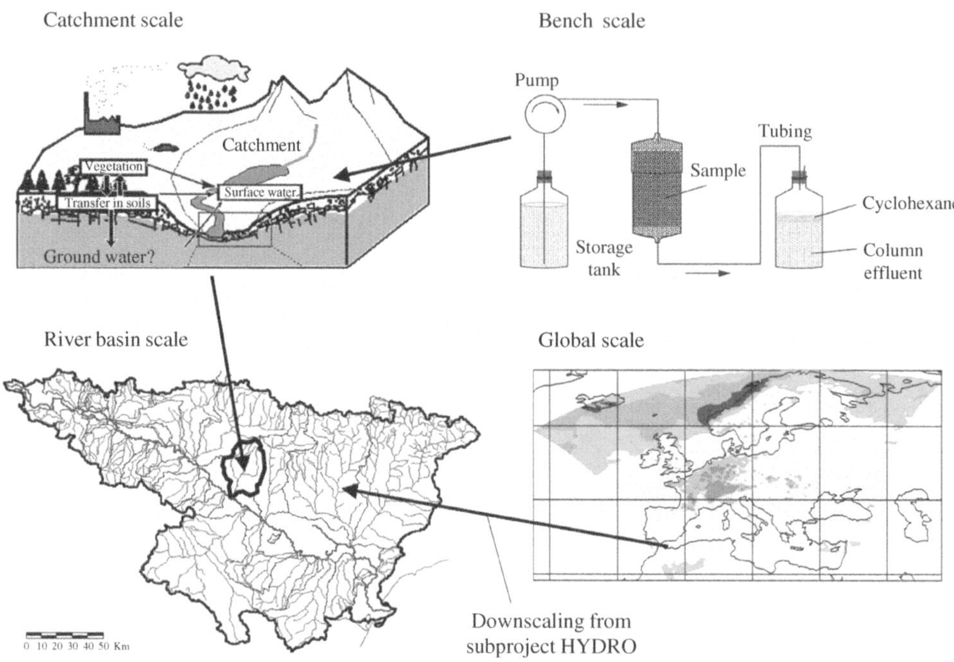

Figure 8.1 Sketch of the different scales addressed in COMPUTE.

Modelling at the bench and hill slope scales

This part of the project deals with modelling of soil water dynamics at the hill slope scale by means of a one-dimensional vertical dual-continuum model (S1D), which is capable of describing the highly transient nature of near surface hydrological processes, and a one-dimensional shallow subsurface flow model (HYPO), both operating at the local (hill slope) scale. In order to model accurately such processes, which are chiefly controlled by preferential pathways, substantial improvements have been introduced in the software S1D, developed by the Czech Technical University in Prague, and MARTHE, developed by CTU and BRGM, respectively. These two modelling framework, have been used to predict the mobility of passive tracers in heterogeneous soil profiles with preferential flowpaths in a few case studies. One of the results of the analysis is the need to consider the appropriate time resolution of rainfall and evapotranspiration intensities to predict the mobility of contaminants in macroporous soils. This is illustrated in Figure 8.2, which shows that the preferential pathways are actively contributing to solute transport only for short periods of time during the major rainstorms. During these rapid transport episodes, the traces moves quickly along preferential subsurface pathways along the hillslope to the river valley.

Furthermore, we studied the effect of particle facilitated transport of strongly sorbed contaminants through macropores by means of the concept of domain specific sorption applied to a dual-continuum system. This concept allows to take into account the mobility of small soil particles in macropores. The particles may carry a substantial amount of adsorbed contaminants (as experimentally confirmed for cadmium). It was shown that the particle mobility can be incorporated in the dual-continuum model by decreasing the sorption distribution coefficient for the macropore domain according to the results of a simple modified batch test (Vogel *et al.*, 2007).

The dual-continuum modeling approach was also applied to study the movement of stable oxygen isotope ^{18}O in a hillslope soil. The isotope has been monitored in precipitation, soil water and subsurface hillslope discharge. Natural fluctuations of ^{18}O signature in precipitation were utilized to detect and analyze preferential flow effects at the hillslope scale. The simulation results confirm the importance of the preferential pathways for the formation of subsurface runoff (Figure 8.3). The simulation also illustrates the possible mechanism of the transformation of ^{18}O input signal, associated with precipitation, into the observed ^{18}O content fluctuations in shallow subsurface runoff (more details in project deliverable C2.19, see www.eu-aquaterra.de). The results of this analysis are an important step forward in the understanding of the main mechanisms of runoff generation at the hillslope scale.

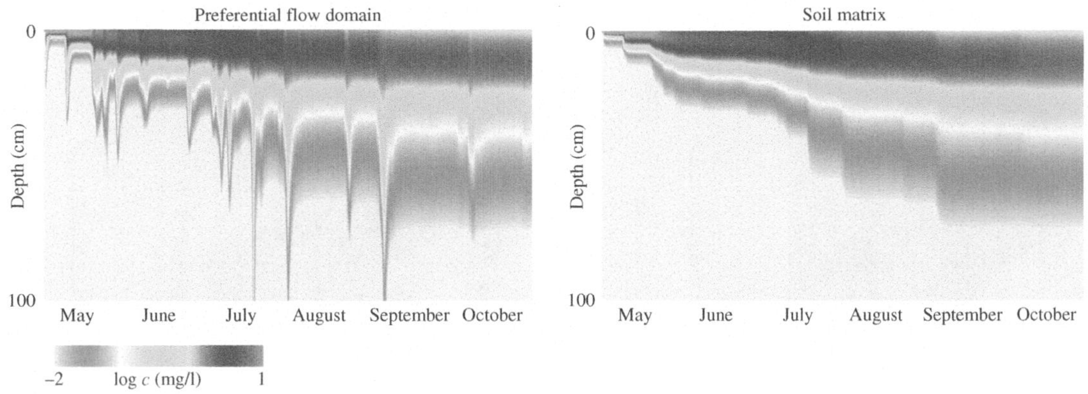

Figure 8.2 Distribution of solute concentration in a macropore soil modeled as a dual-continuum system.

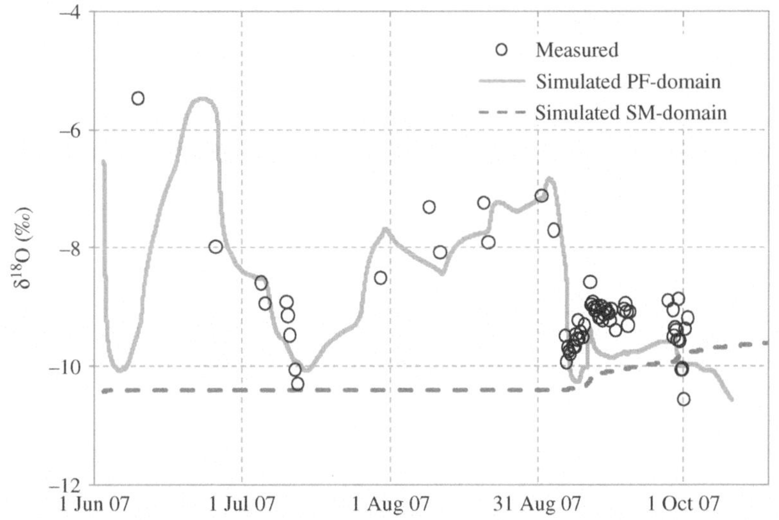

Figure 8.3 Comparison of simulated and observed ^{18}O contents in hillslope discharge (‰ VSMOW). The isotope content was measured in water collected from tipping bucket fluxmeters. The PF-domain contents represent simulated discharge from preferential pathways, while the SM-domain contents are associated with the simulated soil matrix flux.

Modelling at the catchment scale

At this intermediate scale we developed and validated distributed, physically-based models in order to describe hydrological flows (both streamflow and subsurface flow) and the fate of contaminants at the catchment scale. These developments lead to several applications to selected catchments with problems of general interest. For example, the University of Tuebingen applied, in collaboration with TNO, the GeoSys/RockFlow scientific software environment to the Meuse catchment in the Beerze Reusel area. The peculiar morphological nature of the Meuse catchment requires a model capable of coupling surface to subsurface flows over a large area. In order to address properly this issue an accurate and computationally efficient unsaturated zone module has been developed within the framework of a regional hydro(geo)logical soil model (RHSM) currently implemented into the software GeoSys/RockFlow (Kolditz *et al.*, 2008). In principle, a very fine grid is needed in order to properly model flow in the unsaturated zone by solving the Richards equation. This leads to an exceedingly large computational burden when applied to the regional scale. Therefore, we introduced a novel upscaling technique that combines few highly resolved vertical flow profiles deemed representative of areas with similar soil characteristics, which are called influential areas. In doing that only a set of 1-D very detailed flow profiles are solved numerically, thereby reducing the computational burden. In a typical application of this conceptual framework at the catchment scale the large number of influence areas – 50,000 in the Beerze-Reusel area – requires parallelization of the code. Using the message-passing interface MPICH Kolditz *et al.* (2007) showed that the considerable savings in computation time make the simulation of regional soil moisture patterns feasible in reasonable time spans (see Figure 8.4).

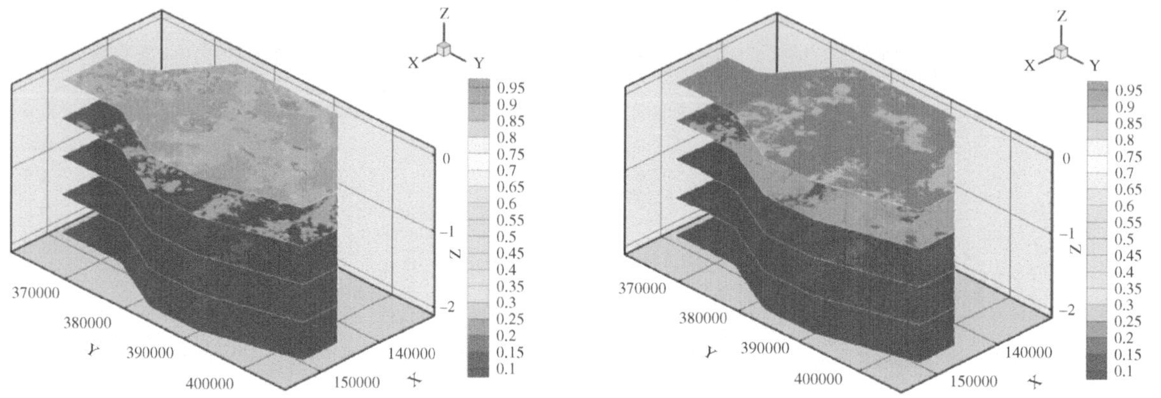

Figure 8.4 Regional soil moisture distribution for the Beerze-Reusel area calcualeted with the parallelised RHSM.

Detailed distributed modelling of the Lerma catchment (total area ~9 km^2), a small catchment within the Ebro river basin, has been conducted by the hydrogeology group of the University of Tübingen during the last year of the AquaTerra project by using the simulation software HydroGeoSphere, which has been developed by the University of Waterloo (Canada). This detailed three-dimensional simulation was possible due to a collaboration with Dr. Jesus Causape from the Instituto Geologico y Minero de España (Geological Survey of Spain), who provided an exhaustive data set on land use change for the catchment. The spatial details for the transition from the hydrologic year 2006 to the hydrologic year 2007 are depicted in Figure 8.5.

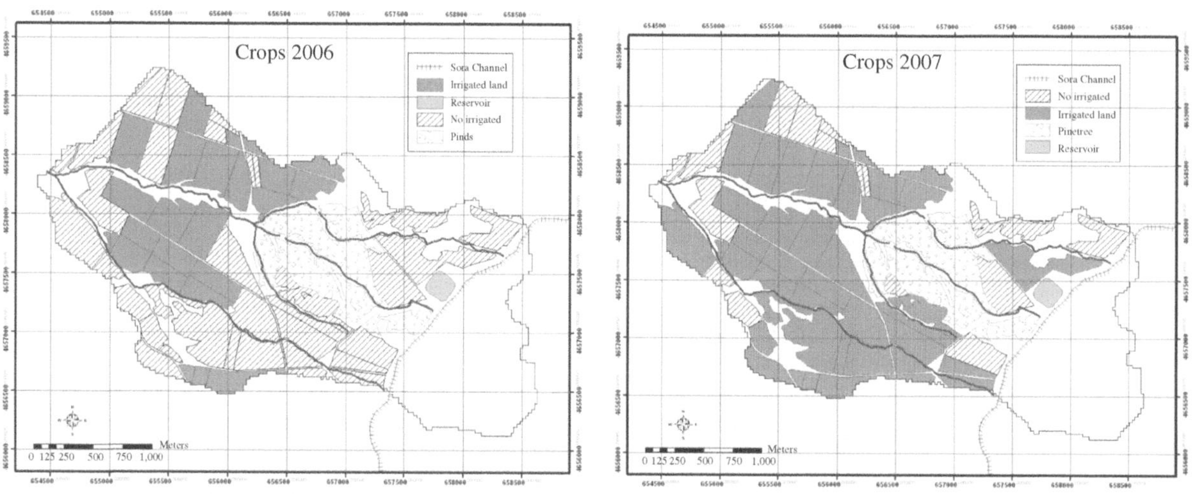

Figure 8.5 The land-use evolution of the Lerma basin for the hydrological years 2006 and 2007.

The modelling study could identify the Lerma basin as a quickly responding catchment in which the hourly characteristics of the rainfall distribution are relatively important for river stage predictions. Nevertheless, after calibrating the hydrologic year 2006 the predictions for the river flow of the hydrologic year 2007 (depicted in Figure 8.6a) are in a good agreement (Coefficient of efficiency = 0.71) with the measurements. This flow model is the prerequisite to study – in a spatially resolved manner – the increase of nitrate loadings (data series shown in Figure 8.6b) at the basin outlet, which were monitored along with the change to irrigation agriculture.

The group of the University of Trento analyzed the impact of catchment topography on hydrological fluxes through detailed water and energy budgets. Simulations have been performed with the software Geotop (Rigon *et al.*, 2006; Bertoldi *et al.*, 2006). Calibration results of a series of simulations in the Brenta catchment (180 km^2), Northeastern Italy, showed the importance of including an adequate reproduction of topography in large-scale land surface models. To facilitate its general use the software has been implemented in a OpenMI environment.

A specific study has been conducted by BRGM in the aquifer feeding the Brèville spring (France). First, a 2D model of the saturated aquifer was built using the MARTHE software to simulate groundwater flow in transient state. After the calibration step, conducted by using 16 observation wells, the model was able to simulate the daily flow and water levels from January 1967 to December 2006.

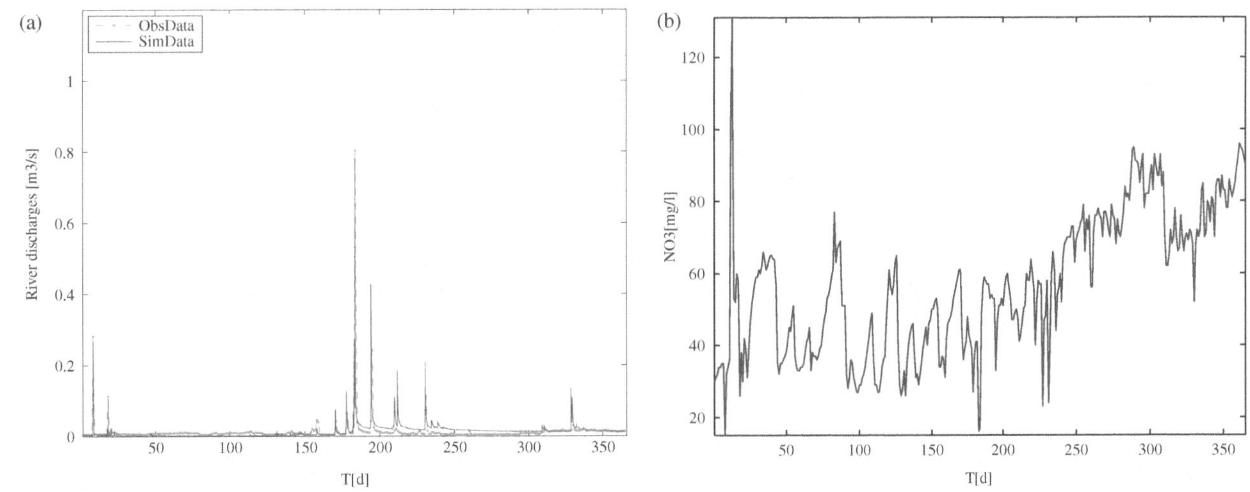

Figure 8.6 (a) Observed vs. computed water discharge for the hydrologic year 2007 (Prediction based on model calibration for the hydrologic year 2006). (b) Daily nitrate concentrations at basin outlet for the hydrologic year 2007.

Furthermore, the concentration of Atrazine at the spring was simulated by applying the software Agriflux to the soil layer. Leaching computed by Agriflux was used as input for the lump 1-D column model (called Nash) of the unsaturated zone. About 72 of these columns are built up to simulate the diversity of pesticide application and the heterogeneity of soil. Output (flow and mass fluxes of pesticide) are then applied into 72 zones at the top of the groundwater saturated zone model (Figure 8.7). Results at the spring, which can be considered as the integrator of the whole aquifer system within the boundaries of the spring catchment, is close to observed data, except during a few high rainfall events in 2000–2001, probably because the computation presented did not take into account preferential flows. Local behaviour of hydrological quantities, such as hydraulic head at the piezometers, is well represented by the model, but the results are less satisfactorily for pesticide transport.

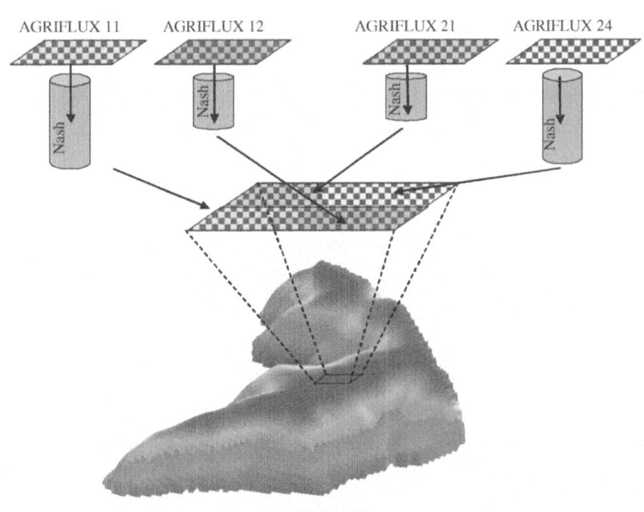

Figure 8.7 Methodology of computing flow and transport in the Brévilles Basin.

The model chain (Agriflux+Nash+Marthe) is used to simulate prediction of pesticide evolution at the spring under various climatic scenarios computed within HYDRO 1 work package (Figure 8.8).

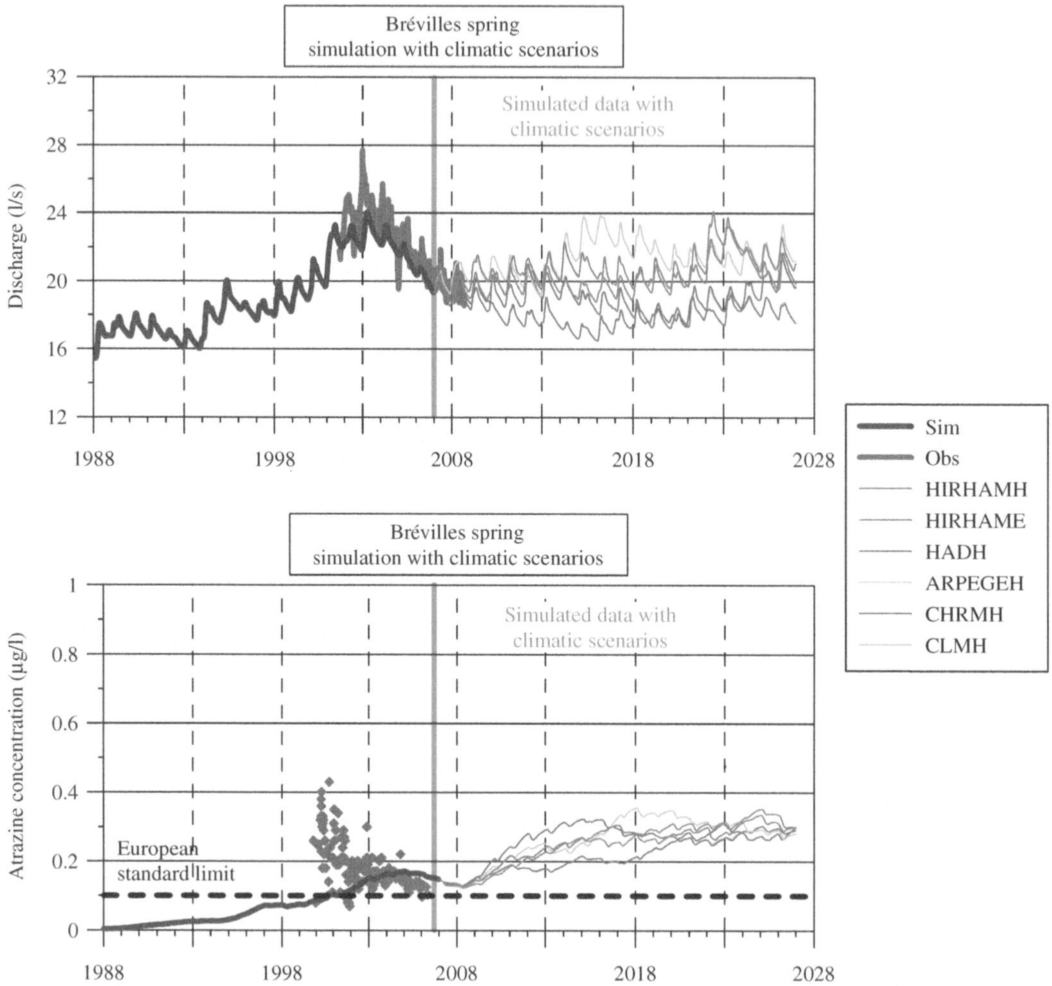

Figure 8.8 Prediction of flow and atrazine concentration at the Brévilles spring.

Modelling at the river basin scale

In order to develop new hydrological models for modelling flow and solute transport at the catchment to river basin scales we introduced a theoretical framework founded on the concept of travel time. Hereby a class of models can be developed with the objective of obtaining the higher prognostic capabilities with the minimum level of complexity compatible with the accurate description of the mechanisms controlling runoff production and solute transport at the basin scale. One important consequence of this paradigm is that the number of parameters requiring calibration is significantly reduced with respect to a typical distributed modelling approach. Model flexibility is granted by a modular organization and by subdividing the catchment in geomorphological units composed of sub-catchments and channels connecting them to the catchment's outlet. Therefore, the river basin is subdivided into sub-catchments feeding the channels which compose the river network. Each sub-catchment represents a geomorphological unit with uniform properties where water and energy balances are applied in a simplified form in order to obtain a parsimonious, yet realistic, representation of the hydrological cycle and interaction with contaminants migration. River scale processes are then simulated along the river network taking into account the contribution from each subcatchment. This modelling framework has been implemented into the software GEOTransf (Majone *et al.*, 2005; Bellin *et al.*, 2010) and used to model streamflow at the Gallego catchment (Spain). The group of the University of Padova has simulated nitrate transport at the Dese catchment (Italy) (Botter *et al.*, 2005).

The group of the University of Trento applied the GEOTransf approach to model streamflow at several control sections along the main course of the Gallego river, one of the most important tributaries of the Ebro (see Figure 8.9). The Gallego river runs from North to South from the Pyrenees to Zaragoza where it flows into the Ebro. In Zaragoza, at the conjunction with the Ebro, the Gallego catchment is of 3690 km^2. A distinct characteristic of this catchment is a strong climatic trend from South to North with annual mean precipitation ranging from of 400 mm in the Zaragoza area to 2000 mm in the northern part of the Pyrenees. Furthermore, in the lower part of the catchment streamflow is severely reduced by water withdrawals for irrigation, while in the upper part the natural streamflow is perturbed by several reservoirs for hydroelectrical power production. These characteristics pose severe challenges to a hydrological model.

Figure 8.9 The Gallego River flows into the Ebro at Zaragoza (left), and a satellite view of the Gallego catchment (right).

An integrated modelling framework that integrates bio-physical processes (hydrology and salinisation of water and soils) with decision making and socio-economic processes (how water is allocated and used by different users, in particular farmers) has been developed and applied to the Gallego. This modelling platform has been used to assess environmental (salinity, water quantity) and socio-economic impacts of several scenarios concerning climate change and water use. The application focuses on irrigated agriculture, which is both a driver (because of its high demand of water resources) and an impacted economic sector in case of reduction of water resources availability. The activities conducted are a result of an interdisciplinary work that combines and integrates input from COMPUTE (University of Trento), INTEGRATOR (BRGM, Wageningen University, Acteon), FLUX (BRGM) and HYDRO (University of Newcastle). The following driving factors affecting salinity and water resources of the Gallego river basin were identified and then compiled in a "global change scenario" that was then simulated by using the above modelling framework:

- Climate change potentially increases temperature and reduces precipitation. This affects water availability and could affect water salinity too.
- Modernisation of agriculture, for example through the implementation of more efficient irrigation techniques, affects the efficiency of water use by agriculture and the volume of salt leaching in return flows.
- An increase of water storing capacities that could enable the extension of irrigated agricultural land in and outside the basin.

One of the main achievements of this study has been the ability of the model to predict with good accuracy river streamflows both in the upper and lower portions of the catchment provided that good quality data are available for meteorological variables (mainly precipitation and temperature), and water uses. In particular, GEOTransf showed good flexibility in handling heavily disturbed catchments allowing to separate climatic and anthropogenic effects due to water withdrawals and transfers (see Figure 8.10).

Additionally the development of the integrated modelling framework provides a practical tool capable of simulating future climate and agro-economic scenarios on the hydro-economic system. Among others, the following aspects of this activity are worth mentioning for their socio-economical relevance: (i) the methodological construction of the modelling framework which couples effectively physical and socio-economical processes and (ii) the level of impact of scenarios on the system. Climate change, modernization of irrigation techniques and dam extension appeared to induce different levels of changes. The main result is that modernization of the irrigation and a larger storing capacity of the reservoirs are effective in reducing the adverse effects of the foreseen climate change scenarios. Another interesting result is that while modernization of irrigation techniques is positive because increases the profitability of the farms, it increases also salt concentration in waters, therefore impacting negatively the environment with a long term negative feedback on agriculture activities and their profitability.

UT carried out hydrological simulations in order to assess the likely impacts of future climate scenarios on water resources for two selected catchments in the Ebro River basin. The selected catchments are evidenced with red circles in Figure 8.11. Parameters of the hydrological model were obtained by calibration to streamflow in the IPCC control period

1961–1990, while the meteorological forcing for the future scenarios (from 2071 to 2100) were provided by HYDRO, as a subset of six different Regional Climatic Models (RCMs) selected from the red set of the PRUDENCE project. Regarding projected climate changes, the six RCMs foresee, in average, a decrease of 10% in the mean annual precipitation and an increase of 4.8°C in the mean annual temperature.

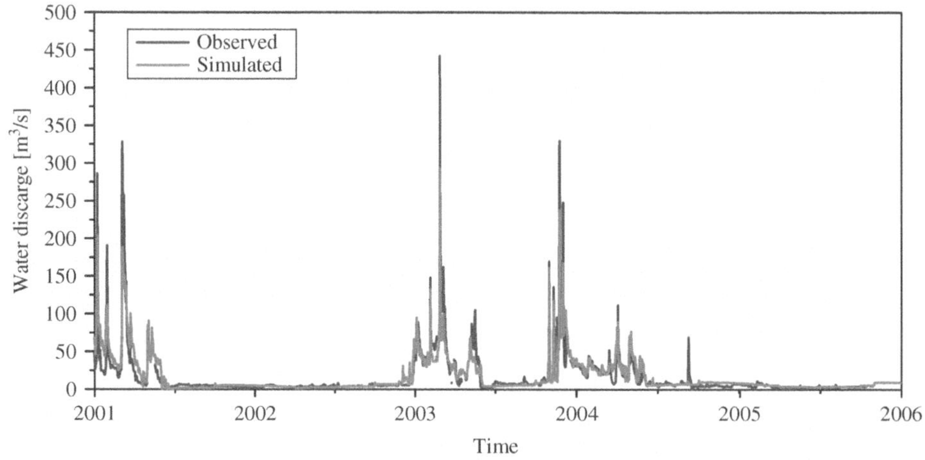

Figure 8.10 Observed (black line) versus simulated (grey line) streamflows at the streamflow gauge station of Zaragoza for the period 2001–2005.

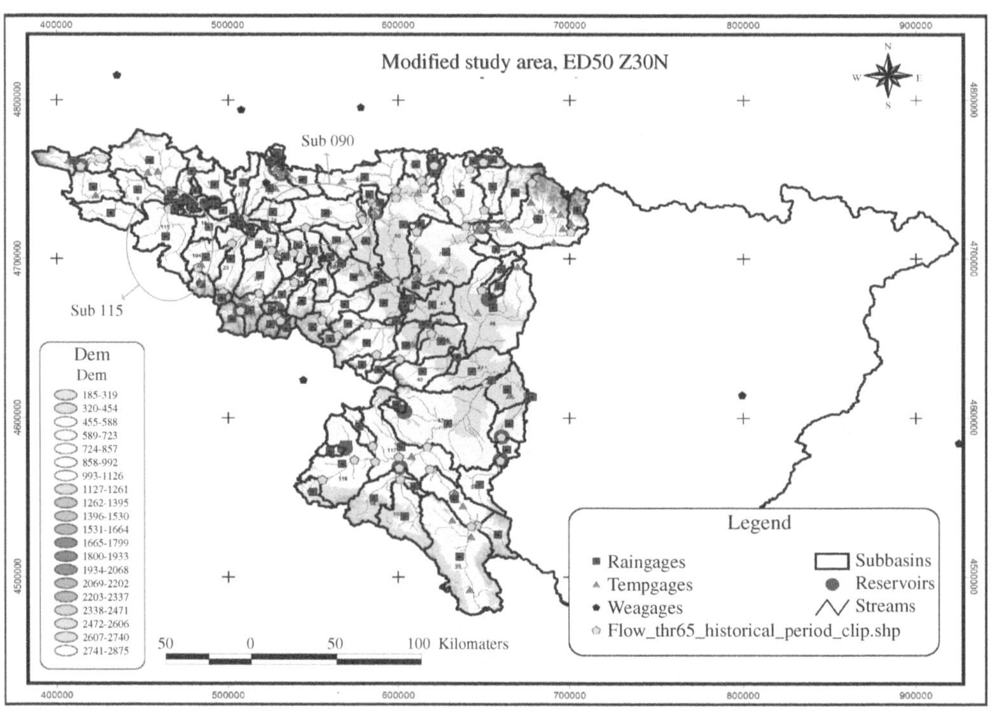

Figure 8.11 Study area, digital elevation model, sub-basins used for the hydrological simulations, together with the location of the gauging stations of precipitation, temperature and streamflow. Grey circles indicate the location of the two catchments selected for the hydrological simulations.

Notice that projected climate changes show large spatial and seasonal variations, which may have important effects on the water availability of the Ebro River basin. In fact, projected decreases in mean seasonal precipitation are concentrated during seasons characterized by intense irrigation activities, and they are only partially compensated by a smaller increase during the winter season. Figure 8.12 shows an example the spatial distribution of the anomalies in the mean annual precipitation and temperature, computed for each one of the six RCMs with respect to the control period 1961–1990.

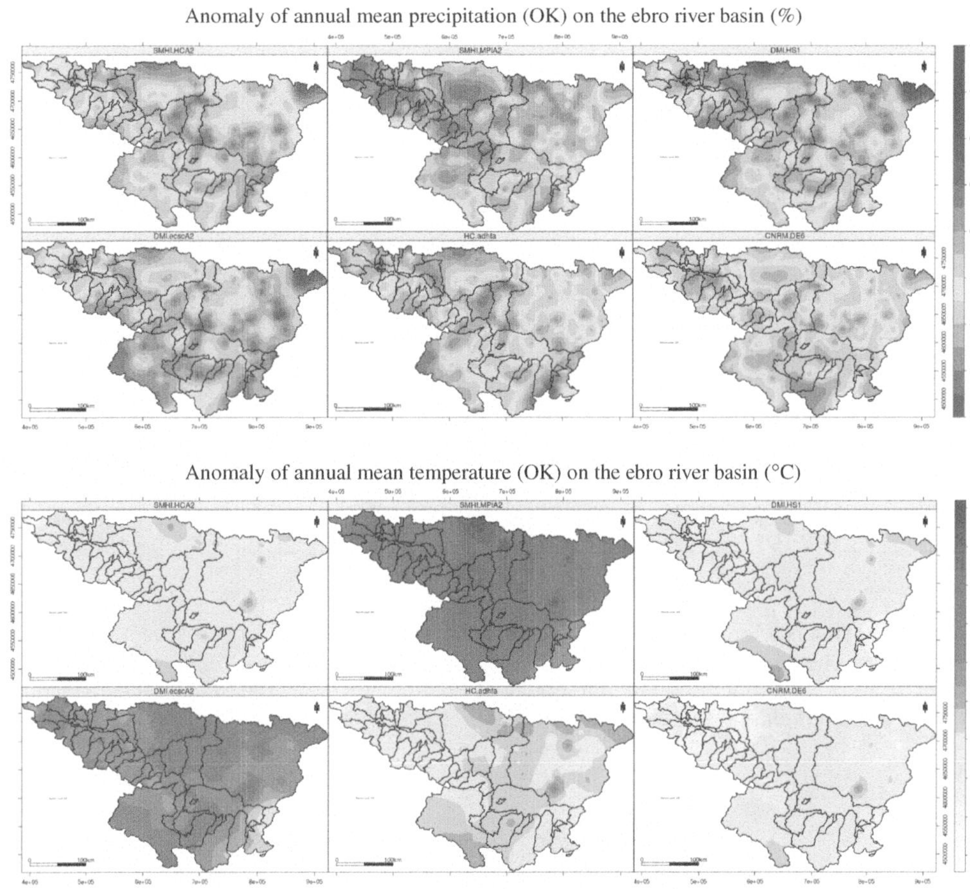

Figure 8.12 Spatial distribution of the anomalies of annual mean precipitation and temperature fields for the Ebro river basin computed for each of the six RCMs adopted in the study with respect to the control period 1961–1990. Data on future scenarios are provided by HYDRO.

In the future scenarios a general decrease is projected for streamflows, resulting from the combined effect of a drier and warmer climate, with respect to the control period 1961–1990. Expected changes in streamflows are summarized in the seasonal daily flow duration curves (FDCs) depicted in Figure 8.13. Changes in streamflows were in general agreement with projections for precipitation and temperature fields, since projected daily streamflows derived from the six climate scenarios were smaller than those observed during the control period.

The group of the University of Padova applied a similar modelling framework to the simulations of the nitrate loads in Dese Catchment (see Figure 8.14), which is a (roughly) 90 km^2 large, flat basin discharging into the Venice lagoon, North-Eastern Italy (Botter *et al.*, 2006, Rinaldo *et al.*, 2006a,b). It is an elongated basin whose mainstream is about 35 km long, geomorphically characterized by groundwater-fed headwaters (where drainage density is low) and mostly left tributaries, draining both highly urbanized and/or extensive agricultural areas, owing to a gentle oblique slope of the Venetian plane. The predictive capabilities of the hydrological model was tested through a long-term simulation of the system's response to meteorological forcings from April 1999 to May 2000 (Figure 8.15). Non-stationary conditions in the model include seasonal variability of parameters. Figure 8.15 shows the substantial agreement of the modelled and the measured water discharge during the considered period, both during major rainfall events (which are reported in the insets) and within dry periods when tidal fluctuations significantly affect flows and hydrometric levels at the outlet. We thus could show that the carrier can be well modelled on this scale.

Once the flow is correctly reproduced, modeling transport requires the calibration of two additional parameters which lump the reaction kinetic. Figure 8.16 shows a good performance of the modeling framework for the nitrate load at the outlet of the catchment in both calibration and validation periods. As expected, mass flux is better reproduced during the calibration period, but good results are obtained also during the validation period, confirming good predictivity of the model. Figure 8.16 shows the complexity of the carrier hydrologic flowrates and the contrast with the regularity of simultaneous nitrate concentrations, both of which are described well by our tools in predictive mode.

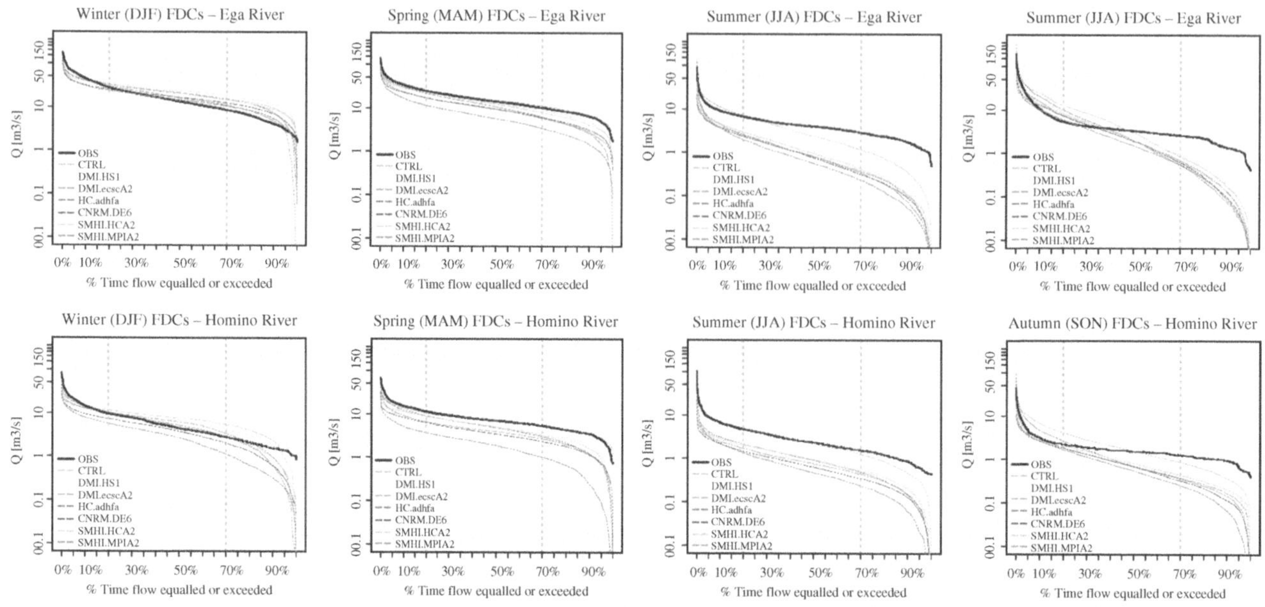

Figure 8.13 29-years seasonal flow duration curves (FDCs) for subcatchments 090 (Ega River) and 115 (Homino River). Black continuous line represents the observed seasonal FDCs during the period 1962–1990, grey dotted line represent the seasonal simulations for the same control period, and coloured dotted lines show the seasonal simulations corresponding to the different climate scenarios during the period 2072–2100.

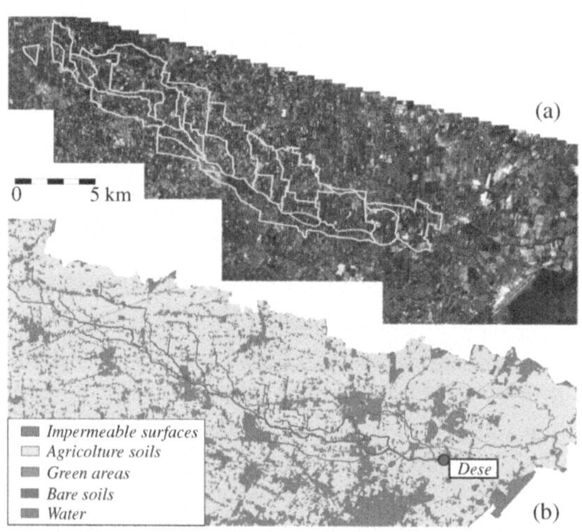

Figure 8.14 (a) Remotely sensed image (IKONOS) of the Dese river basin (provided by Consorzio Venezia Nuova-Magistrato alle Acque di Venezia) and partition into subbasins; (b) Thematic map of land use of the catchment: classes of land use that are relevant to the modelling of water balance processes.

The group of the University of Padova has also designed an hydrocory framework capable of modelling ecological processes on river networks. Coupled models of transport and reaction were developed for: (i) spreading of waterborne diseases (ii) riverine biodiversity and (iii) migration/invasion processes. River Network is modelled as an oriented graph constituted by nodes and edges. The models are assembled by coupling local reaction models at the nodes with a transport model through the edges. Basin-scale transport processes, the core of the AquaTerra project, are a major ecological player because river networks define the ecological corridors along which many crucial hydrologic, biologic and ecologic processes take place. Thus all spatial attributes of the invasion processes depend on the network geometry and topology and on transport properties that can be seen in a unified perspective.

This research group designed a spatially explicit model for the study of the spreading of waterborne diseases. The model explicitly accounts for the role of the river networks in mixing and redistributing the pathogens responsible for the disease. The model has been then particularized for the study of cholera and applied to the study of the spatio-temporal evolution of a cholera epidemic that occurred in the KwaZulu Natal province of South Africa (Bertuzzo *et al.*, 2008). All the necessary epidemiological, demographic and hydrologic data were collected. The epidemic lasted for two year involving 135000 confirmed cholera cases. Figure 8.17 compares the spatial distribution of data with that obtained via simulation. It shows the distribution of the cumulated cholera cases after the first and second epidemic outbreak. Colors are graded via spatial linear interpolation of nodal values. The model proved capable of capturing the spatio-temporal patterns of the disease spreading from the coastal region to the inland disease free region thus highlighting a clear hydrologic control on the spreading of water borne diseases.

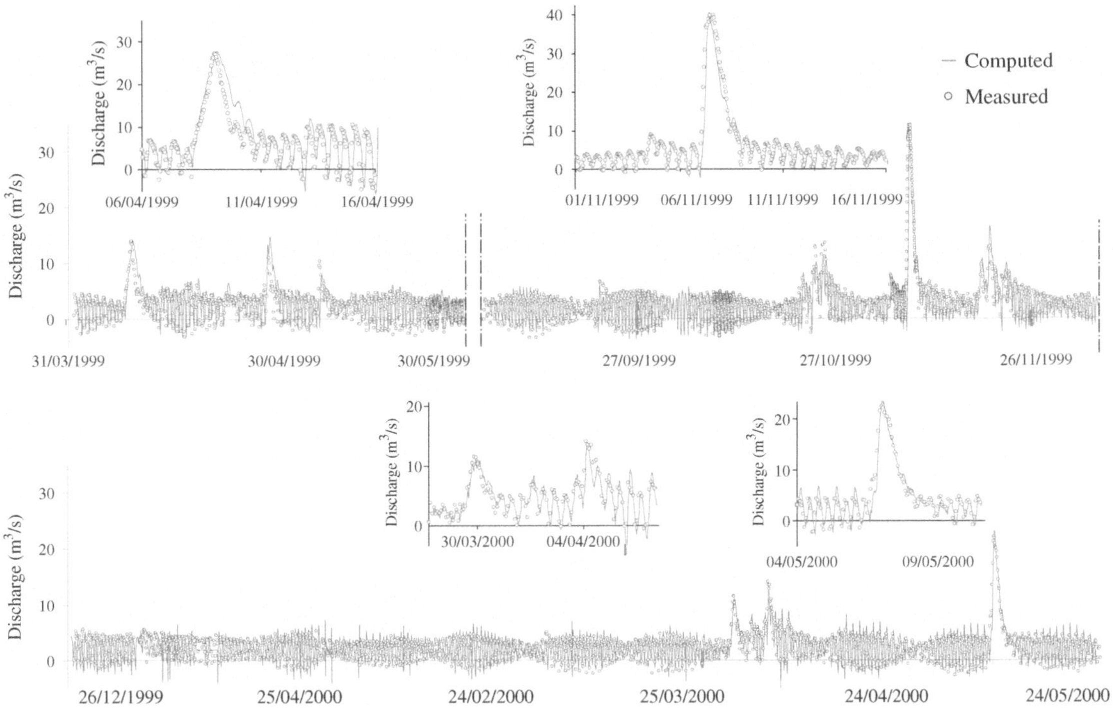

Figure 8.15 Validation of the flow model: long term (one year) simulation of the discharge at the outlet of the Dese Basin: comparison between computed and measured time series.

Successively the research group attempted to show that the river network structure acts as effective template for characterizing spatial attributes of biodiversity, in particular fish biodiversity, and that estimates of average dispersal behavior and habitat capacities, objectively calculated from average runoff production, yield reliable predictions of large scale spatial biodiversity patterns in riverine systems. The model is of structured metacommunity type. The neutral theory of biodiversity (Hubbell, 2001) is implemented in the river system, using its network as the structure of the metacommunity. Each local community in that metacommunity has a different fish habitat capacity that is assumed to be function of the mean annual runoff production. This model has been applied to study the riverine biodiversity of the Mississippi-Missouiri River System (MMRS). The vast extent of the MMRS spans across diverse habitat types operating under varying environmental conditions (e.g., climate hydrologic regime, primary productivity, human disturbance); these habitats are connected to one another by the MMRS river network. Figure 8.18 summarizes the data collected and used in this analysis. The model proved capable of reproducing pattern of local species richness, geographic range and between community similarity (Muneeperakul *et al.*, 2008). The results suggest that parameters characterizing average fish behavior, habitat capacities and connected structure suffice to obtain reliable predictions of large-scale biodiversity patterns.

Software tool box

The capabilities of the software toolbox components have been largely extended. New modelling concepts have been successfully implemented in the AquaTerra toolbox and some selected details have been described above. It is important to note that in addition to the new modelling approaches, the tool box has been applied to almost all AquaTerra river basins (see Table 8.1).

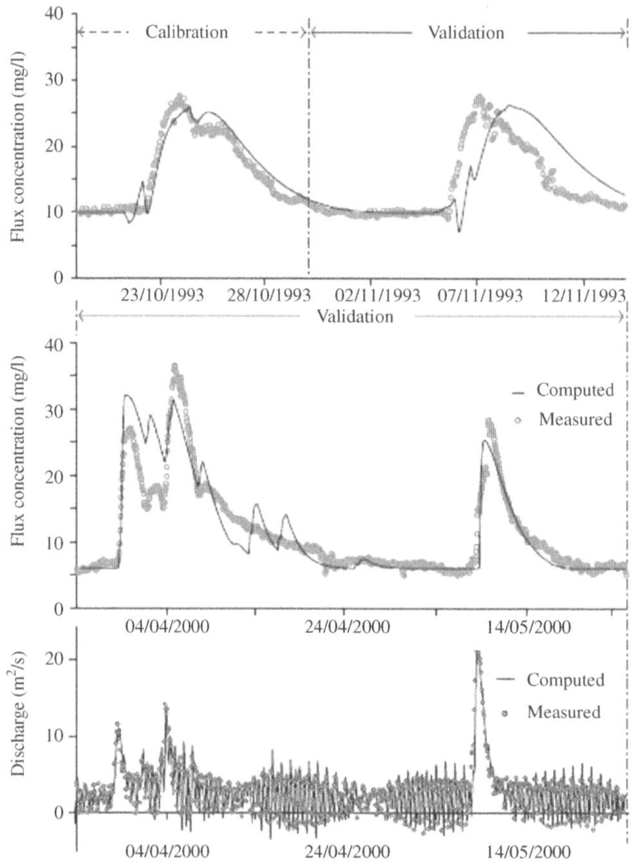

Figure 8.16 Validation of the single layer transport model: nitrate concentration computed and measured at the gauging station during different time periods: October–November 1993, and April–May 2000. In the lower graph the temporal evolution of the discharge during April–May 2000 is also reported.

IMPLICATIONS AND APPLICATION OF RESULTS

Within COMPUTE we developed a suite of modelling approaches, which in the future can be applied to other case studies and catchments. In particular, the modelling frameworks have been used to study soil water dynamics at several spatial scales, thereby providing:

(i) Realistic estimates of the recharge to the groundwater, which is important to assess groundwater availability and renewal; better ability to quantify water flows is the foundation to enable pollutant transport and flux quantifications.

(ii) Quantitative estimates of the soil-atmosphere feedbacks, which have important influences on microclimate with applications in environmental impact studies (i.e., localization of industries or other large facilities, such as power plants, urban soil waste treatment plants, etc.). In addition, the proposed modelling framework can be used in environmental studies concerning the overall impact on the aquatic ecosystem of contaminants and nutrients releases both from diffused and concentrated origin.

(iii) Quantitative estimates of the impact on water resources of future climatic and water and soil use scenarios. Applications have been performed in the Ebro river basin as a blueprint for future applications in other river basins.

CONTRIBUTIONS AND POLICY RECOMMENDATIONS FOR RIVER BASIN MANAGEMENT

The main recommendation to river managers is that existing measurements networks should be improved by increasing both spatial and temporal resolution. This is particularly true for water quality measurements. The recent development of inexpensive wireless sensor networks (which results from two important technological advances: the availability of new and inexpensive sensors and the possibility to connect them through a wireless network) can help in alleviating investments and operational costs, which hampered the development of ground based sensor networks in the past.

Calibration, feedback and prediction through modelling would then become a reality at much higher resolution and detail. The exercises we conducted within AquaTerra showed that robust and well tested modelling systems may be very useful to envision new management strategies for anticipating and mitigating the adverse effects of climate and soil use changes. Another recommendation is that changes in water use and hydraulic infrastructures (i.e. reservoirs) should confronted with the results of accurate simulations prior implementation. The simulations should involve the processing through hydrological modelling frameworks of the type developed in AquaTerra, of reliable future climate, water and soil use scenarios. This will require including scenario development and study in the process of developing river basin management plans.

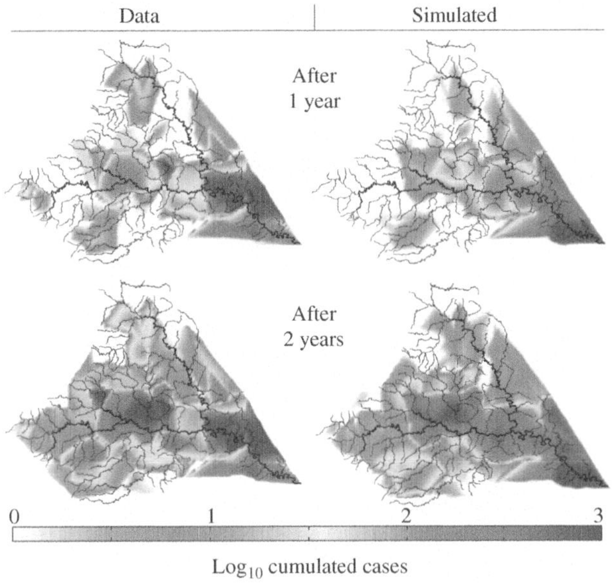

Figure 8.17 Comparison between data and simulated spatial distribution of the cumulated cholera cases after the first and second epidemic outbreak for the Tukhela river basin. Colours are obtained via spatial linear interpolation of the node values.

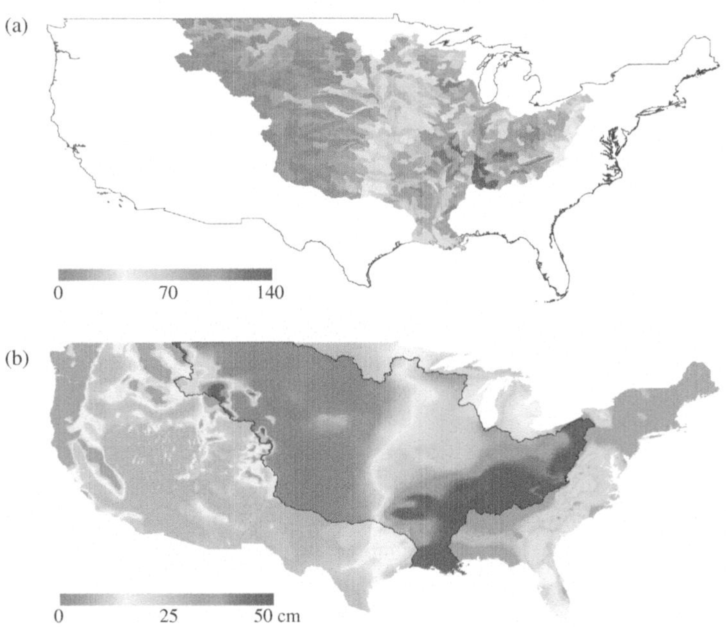

Figure 8.18 The maps of (a) local species richness (LSR), or α-diversity, of the freshwater fish, and (b) the average annual runoff production (AARP) in the Mississippi-Missouri River System (MMRS).

Table 8.1 Overview of the tool box applications to AquaTerra river basins.

Code	Purpose	Scale	References
S1D HYPO	Highly transient near surface hydrological processes; applied in Elbe Basin	Soil profile	TU Prague Vogel *et al.* (2007)
MARTHE	Tracer mobility in heterogeneous soil applied in Bréville Catchment	Soil profile	BRGM
GeoSys	Hydrological processes with high-resolution soil models; applied in Meuse basin (Beerze-Reuzel and Geer Catchments both within Meuse Basin)	Catchment	TUB/UFZ Kolditz *et al.* (2007), Delfs *et al.* (submitted)
Feflow/ Mike11	Hydrodynamic processes; applied in Elbe basin	River branch	DHI-WASY
GeoTop	Hydrological fluxes through detailed water and energy budgets; applied in Brenta catchment	Catchment	UT, Rigon *et al.* (2006), Bertoldi *et al.* (2006)
Geo-Transf	Hydrological models for flow and solute transport based on the travel time concept; applied in Gallego catchment (Ebro Basin), Dese river catchment	River basin	UT, Bellin *et al.* (2010), UP, Botter *et al.* (2005)

Locally/nationally

At the local scale, detailed studies on the impact of climate and soil use changes have been completed. These studies encompass scales ranging from a few square kilometres (Brévilles and Lerma catchments) to several thousand kilometres (Gallego and Ebro river basins).

Modelling results at Brévilles indicates that pesticide contamination by atrazine will take decades to come back to acceptable quality. Although the calibration of the modelling could be improved, these results confirm field observation (Gutierrez et Baran, 2009) and may be considered as a relevant indication for decision makers in the frame of the European Directive. However one should avoid assimilation of atrazine, a very persistant herbicide banned in France since 2003, and other kind of pesticides. The analysis has also shown that diffuse contamination by pesticide is not only depending on soil characteristics and hydrogeological settings but also on the type of pesticide.

The modelling activity conducted in the Gallego and Ebro river basins suggest that, under climate change scenario, future water resources will not be sufficient to supply the foreseen maximum irrigation demand. With limited irrigation water, farmers will be limited in their choice of crops, crops will suffer water-stress and agricultural yield could decrease considerably, and with it, the income of farmers. Currently, irrigation in the Gállego is predominantly based on inefficient flooding methods with only a small percentage based on sprinkler, sprinkler with pump, or drip irrigation systems. Significant volumes of water could therefore be conserved by increasing the on-farm irrigation efficiency. In order to cope with this challenge, a National Irrigation Plan ('Plan National de Regadios', 329/2002) has been launched in 2002. The objective of this plan is to restructure, modernise and consolidate agriculture in Spain to reduce the environmental impacts of irrigation. Although this plan encourages the adoption of more efficient irrigation systems, the shift to new technologies may be slow. It may be possible to create additional storage volumes but building new reservoirs, or increasing the volume of existing ones, is a controversial issue and may create conflicts (e.g., the controversial proposal to build a dam above the Ardisa reservoir in the Gallego river basin). The modelling framework developed within AquaTerra may help in developing reliable future scenarios in order to verify in a consistent manner the multifaced effects of changes in water management and construction of new reservoirs, thereby helping in making informed decisions. An important outcome of this scenario oriented decision making process is the possibility of confronting alternative solutions against possible future scenarios, thus gaining more confidence in the implemented solution.

In a European context

At the European level the research activity of AquaTerra highlighted the need for improving monitoring systems of water quality across Europe. For most modelling activities future better availability of data is central for success. In most cases, relevant contaminant loads are carried by rivers during high flow events while monitoring of most contaminants is at the best organized on monthly basis with strategies that lead to preferential sampling during low flow periods when the concentration levels are assumed to be the highest. Although this may be the case, the total load is controlled by the mass flux (i.e., the product of solute concentration by the water flow), which is typically higher during short time high flow periods. This is particularly true in agricultural areas where high intensity rainfalls mobilize the nutrients stored in the upper soil strata. This calls for the development and installation of new automatic high time resolution monitoring systems capable to increase the sampling frequency from once a month, typical of the manual systems, to daily or hourly sampling.

Another result with European to global relevance is the need of increasing the spatial density of meteorological measurements in mountain areas across Europe, possibly through a joint effort to install new generation of low cost sensors connected through a wireless sensors network to the existing network of traditional (and typically more accurate) meteorological stations. This is needed because in complex terrain the spatial variability of the climatic forcing may not be captured by the existing gauging stations, which samples preferentially at lower elevations within the valleys.

Besides answering specific local questions raised by river managers, we developed a suite of modelling approaches, which in the future can be applied to other case studies as support tools to the management decisions. These modelling platforms have enabled to combine different change scenarios in order to simulate a possible vision of the future by integrating combined effects of changes. This image and future state assessment, even if uncertain, can assist decision makers in river basin planning by providing them a view of the basin "with actual policies and trends". It can provide a basis to discuss measures to be implemented and identify possible negative trends. Within WFD framework such a tool can be interesting to build the possible state of a basin in 2015 and to assess risks of water bodies. In a participative framework that would explore the adaptation of system to climate change it is possible to use these tools to assess the effect of policies prior their implementation. This could be interesting for decision makers, because it delivers them an indication of possible impacts of a policy. Again, this fits within the WFD which asks member states to select and implement programs of measures that should enable to reach good status by 2015.

OPEN & NEW QUESTIONS FROM COMPUTE

A tighter connection between spatial and temporal density in monitoring systems is one of the important open questions about high resolution data availability that have been highlighted by the project. This has been shown clearly by both distributed and semi-distributed modelling approaches where, particularly in mountain terrains, the available spatial density of measurements is often inadequate to properly reproduce spatial variability of climatic forcing. In addition to that, new monitoring sensors should be developed in order to better characterize the time distribution of contaminant and nutrient loads. New and more accurate methodologies for measuring soil water dynamics across spatial scales is another important open question that – if solved – would open new possibilities in modelling the terrestrial part of the hydrological cycle. Further developments are needed in the understanding of the relationship between climate change soil and water use at the river basin scale. AquaTerra contributed with a robust modelling framework and conducted some interesting analyses within the Ebro river system. However, further developments are expected in this field, which is in its infancy, and new development are expected from the better integration between climate and hydrological modelling.

REFERENCES

Bellin, A., Majone, B., Cainelli, O., Alberici, D. and Bertagnoli, A. (2010) GEOTRASNF: A Continuous hydrological model for water resources management in disturbed catchments, in preparation for Hydrological Processes.

Bertoldi, G., Rigon, R. and Over, T.M. (2006) Impact of watershed geomorphic characteristics on the energy and water budgets. *J. Hydrometeor.* **7**, 389–403.

Bertuzzo, E., Azaele, S., Maritan, A., Gatto, M., Rodriguez-Iturbe, I. and Rinaldo, A. (2008) On the space-time evolution of a cholera epidemic. *Water Resour. Res.* **44**, W01424.

Botter, G., Bertuzzo, E., Bellin, A. and Rinaldo, A. (2005) On the Lagrangian formulations of reactive solute transport in the hydrologic response. *Water Resources Research*, W04008, doi:10.1029/2004WR003544.

Botter, G., Settin, T., Marani, M. and Rinaldo, A. (2006) A stochastic model of nitrate transport and cycling at basin scale. *Water Resources Research* **42**.

Delfs, J.O., Park, C.-H. and Kolditz, O. (Submitted) Interface concept for Hortonian flow. *Water Resources Research*.

Gutierrez, A. and Baran, N. (2009) Long-term transfer of diffuse pollution at catchment scale: respective roles of soil, and the unsaturated and saturated zones (Brévilles, France). *Journal of Hydrology* **369**, 381–391. doi:10.1016/j.jhydrol.2009.02905

Hubbell, S.P. (2001) The Unified Neutral Theory of Biodiversity and Biogeography, Princeton University Press.

Kolditz, O., Delfs, J.-O., Bürger, C., Beinhorn, M. and Park, C.-H. (2008) Numerical analysis of coupled hydrosystems based on an object-oriented compartment approach. *Journal of Hydroinformatics* **10**(3), 227–244.

Kolditz, O., Du, Y., Bürger, C., Delfs, J., Kuntz, D., Beinhorn, M., Hess, M., Wang, W., van der Grift, B. and te Stroet, C. (2007) Development of a regional hydrologic soil model and application to the Beerze-Reuzel drainage basin. *Environmental Pollution* **148**, 855–866.

Majone, B., Bertagnoli, A., Bellin, A. and Rinaldo, A. (2005) GEOTRANSF: a continuous non-linear hydrological model. *Eos Trans. AGU*, **86**(52), Fall Meet. Suppl., Abstract H23C-1441.

Muneepeerakul, R., Bertuzzo, E., Lynch, H. Fagan, W.B., Rinaldo, A. and Rodriguez-Iturbe, I. (2008) Neutral metacommunity models predict fish diversity patterns in Mississippi-Missouri basin. *Nature* **453**, 71–92.

Rigon, R., Bertoldi, G. and Over, T.M. (2006) GEOtop: A distributed hydrological model with coupled water and energy budgets. *J. Hydrometeor.* **7**, 371–388.

Rinaldo, A., Botter, G., Bertuzzo, E., Uccelli, A., Settin, T. and Marani, M. (2006a) Transport at basin scales: 2. Applications. *Hydrology and Earth System Sciences* **10**, 31–48.

Rinaldo, A., Botter, G., Bertuzzo, E., Uccelli, A., Settin, T. and Marani, M. (2006b) Transport at basin scales: 1. Theoretical framework. *Hydrology and Earth System Sciences* **10**, 19–29.

Vogel, T., Lichner, L., Dusek, J. and Cipakova, A. (2007) Dual-continuum analysis of a cadmium tracer field experiment. *Journal of Contaminant Hydrology* **92**, 50–65.

Chapter 9

INTEGRATOR – Integration of knowledge

C. Merly, A. S. Chapman, E. Ansink, A. Bardos, P. Bardos, M. Bouzit, J. Brils, N. Graveline, C. Herivaux and D. Darmendrail

KEY OBJECTIVES

INTEGRATOR aimed to combine the knowledge developed in the AquaTerra project with stakeholders' needs, through effective merging of natural science with economic and social science, involving practitioners and stakeholders identified in the EUPOL and BASIN sub-projects. It developed innovative approaches and conceptual tools to assist the decision-making process with a focus on stakeholders and decision makers.

In order to achieve these objectives, INTEGRATOR was split into three workpackages, which were undertaken sequentially:

WP INTEGRATOR 1 (I1)
The aim of WP I1 was to develop a conceptual integrated model representing the interactions between economic activities and the quality of soil and water resources at the river basin scale. This model was established for four large scale European river basins (the Danube, the Ebro, the Elbe and the Meuse). The integrated model aimed to provide a framework to characterise and assess direct impacts (e.g., those involving physico-chemical processes) and indirect impacts (e.g., those involving economic and social processes) on river basin quality.

WP INTEGRATOR 2 (I2)
The aim of WP I2 was to implement the conceptual model through the identification of the areas and issues to be addressed at the local level and the relevant driving forces and the identification of potential trends for each driving force in order to simulate the effects of policies in the medium and long term. The implementation intended to develop specific decision-making models in three river basins through modelling economic sectors having an impact on soil-water resources and sectors depending on soil-water resources and to validate the conceptual model cases.

WP INTEGRATOR 3 (I3)
The initial objective of WP I3 was to design a set of generic conceptual models of decision making processes based on the synthesis of the different basin conceptual models developed in I1. In addition, I3 aimed to develop a methodological framework for stakeholder involvement and public consultation and to integrate scientific and economic issues in order to make the outlines of a possible future decision support tool (to be developed outside AquaTerra).

Through the development of INTEGRATOR 1 and 2 and the overall development of AquaTerra, it was concluded that there is no one generic model suitable for all sites and all environmental issues. Therefore, the objectives and the approach taken in I3 were refined and adapted as follows:

- Model(s) of the interaction between socio-economic activities and water-soil resources through the development of socio-economic analyses were developed for selected case studies;
- A pictorial web-based representation of the data available in AquaTerra was developed to make the information available better suited to informing decision making processes within and beyond the AquaTerra river basins;
- Delivery of major scientific results obtained in the different SPs to selected operational decision making levels that are the final end-users and;
- Projects relevant to AquaTerra, which may present the opportunity for collaboration and cooperation at a European level were identified.

KEY ACHIEVEMENTS

INTEGRATOR outputs provide a means of supporting decision-making in river basin management through conceptual models combining socio-economic and bio-physical data and the presentation of integrated information oriented towards river basin management issues and more general stakeholder demands. INTEGRATOR results and methodologies were based on strong stakeholder involvement.

INTEGRATOR 1 – Evaluation and characterisation of stakeholder demand for a conceptual model of interaction between the society and soil-water resources

The first integrator sub project generated baseline information about specific river basins (Ebro, Danube, Meuse and Elbe), including policy and management attitudes using a combination of original research and initial consultations with relevant stakeholders.

The integrated approach proposed by INTEGRATOR 1 was based on the generic DPSIR (Driver – Pressure – State – Impact – Response) framework, which has been widely used by the European Environment Agency (EEA) and the European Commission (EC). Using the DPSI(R) framework, the specific soil-sediment-water system components were characterised in the four principal AquaTerra river basins (project deliverables I1.1c to I1.1f, see www.eu-aquaterra.de). The common key drivers identified were demography, environmental and agricultural policy, the nature of industry within each basin and its level of development, agriculture and climate change. These drivers have been identified as having a major impact on water use and water resources, and thus have a major impact on the river basin system as a whole. Pressures and States were analysed up to the present time through physical indicators such as suspended solids and heavy metal concentrations in surface water, groundwater, sediment and soil components. The main impacts on ecosystems and human activities were also identified for each basin. This characterisation process resulted in the initial description of the key environmental issues at each of the AquaTerra selected river basin, as shown in Table 9.1.

Table 9.1 Summary of the main issues at stake in the four indicated river basins identified with stakeholders participation.

Main issues at stake	Meuse	Elbe	Danube	Ebro
– diffuse (agriculture) & point source pollution of surface water	√	√	√	√
– contaminated sediments & floodplain soils (a.o. heavy metals)	√	√	√	√
– flooding	√	√	√	
– large scale groundwater pollution due to past mining/industry activity	√	√		
– damming/reservoirs & resulting sediment deficits downstream			√	√
– soil erosion			√	√
– water scarcity				√
– salinisation				√

The DPSIR framework was developed for each component of the soil-sediment-water system and was used to describe the key environmental issues in eachriver basin (Figure 9.1). This work was presented in river basin characterisation reports for each basin (project deliverables I1.1c-f) This information was subsequently combined with the opinions of EC decision makers and stakeholders (based on interviews and workshops) as well as current policy targeting river basin management were brought together in (project deliverables I1.2). The results of this activity showed anticipated future changes in climate and economy were liable to be significant and would impact on the river basin as defined by the DPSI(R) framework. The implementation of water policies was an additional source of potentially large effects on water pollution and abstraction. The scale of these changes and their effects are, however, uncertain and show large variations across regions due to variationsin climatic conditions, economic activities and policy structures.

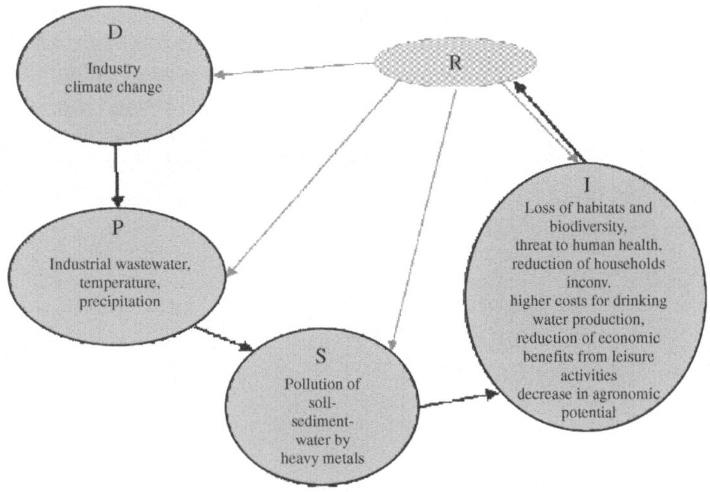

DRIVERS	PRESSURES	STATE	IMPACTS
Industry (metal processing, manufacturing of paints and plastics...) Climate change	Industrial wastewater	Pollution of surface water Pollution of soil/ sediment	Increase in the consumption of bottled water Higher costs for drinking water production Loss of habitats and biodiversity Damage for leisure activities (incl. fishing) and tourism Decrease in agronomic potential of arable land

RESPONSES			
Moving/banning polluting industries	Taxes based on the heavy metals loads	Decontamination of polluted soils/ sediments	Improvement of treatment process for drinking water production (subvention)

Figure 9.1 Development of the DPSIR framework for heavy metal pollution in sediments.

INTEGRATOR 2 – Implementation in selected case studies

While INTEGRATOR 1 assessed the likely sources of significant wider impacts on the state of river basins (project deliverable I1.2), the impact of global change on economic activities in the four European river basins was assessed by INTEGRATOR 2 (project deliverable I2.2). One of the conclusions was that, although considerable progress is being made in the management of pollution in the four river basins, economic and climate changes provide additional pressures. The main changes identified in the case study river basins were industrial growth, agricultural policy change, hydro-morphological changes, reduced stream flow and the increased risk of floods and droughts.

Participatory workshops involving policy makers from regional and national governments, Water Agencies, River Basin Commissions, water and wastewater operators, representatives of agriculture, industry, tourism and scientists were organised in the relevant case study areas (see Table 9.2). The aim of these workshops was to complete and validate the table of key environmental issues identified within each basin through the characterisation reports and to discuss the potential future changes in the drivers affecting the system. Key results of these workshops are summarised in project deliverables I2.3a–d for the Meuse, Elbe, Ebro and Danube workshops respectively.

On the basis of the results obtained in INTEGRATOR 1 and INTEGRATOR 2, five case studies were selected for the development of conceptual models integrating economic analysis with the physical processes assessed within AquaTerra (project deliverable I2.5). These case studies are summarised in Table 9.3.

The Ebro and Geer catchments and the Kempen Region were selected to achieve integration with the objective of supporting decision making at the case study level (see also project deliverable I2.4).

Table 9.2 List of stakeholders' workshops.

River Basin workshops	Date and location	Organised by	Type of stakeholders
Meuse River Basin Workshop	15 September 2005, Liege, Belgium	Brgm, TNO and ISSeP	Scientists, water agencies, policy makers, RB International Commission, Environmental regional authorities, Industrial representatives
Elbe River basin stakeholder workshop	10 November 2005, Leipzig, Germany	TU Hamburg-Harburg and UFZ Leipzig Ltd	Scientists, representatives of NGO's, industrial, representatives of approximately local, regional and federal authorities
Danube basin workshop	24–25 March 2006, Budapest, Hungary	Jointly organized by UNESCO, IHP/ISI, SEDAN, AquaTerra, SedNet, in cooperation with ICPDR and VITUKI	Scientists, policy maker
Ebro basin workshop	7 April 2006, Barcelona, Spain	INSTITUT D'ESTUDIS CATALANS, r3 Environmental Technology	Scientists, union representatives, government agencies,
Elbe River basin stakeholder meeting	16 May 2007, Magdeburg, Germany	TNO	Scientists, NGO's representatives, government representatives

Table 9.3 Main socio-economic sub-systems considered in the conceptual models of the INTEGRATOR case studies.

Case study area	Environmental Issue considered	Main external drivers	Socio-economic sectors and sub-systems considered
Geer Catchemnt (Meuse, Walloon region)	Diffuse nitrate pollution in groundwater	Agricultural policy Environmental policy (e.g. Nitrates Directive)	Drinking water sector Household water use Industry (agro-food) Agriculture
Kempen area (Meuse, Flanders and Netherlands)	Soil contamination by heavy metals	Industrial policy Health policy & food safety policy Land planning Environmental legislation	Population (health) Drinking water sector Agriculture Urban development
Central Ebro River Basin (Ebro, Spain)	Soil and surface water degradation because of salinity/sodicity	Agriculture policy Tourism development Population growth Industrial development Water policy (e.g. National Hydrological Plan)	Agriculture Drinking water sector Industry & public equipments Fishery (?)
Krska kotlina aquifer (Danube, Slovenia)	Groundwater pollution by nitrates and pesticides	Population growth Tourism development Agricultural policy Environmental & water policy (UWWT & Nitrates directive)	Agriculture Municipal (household) water service sector Disconnected households Thermal tourism
Meuse river basin (Flanders & Netherlands transboundary part)	Water quality and water quantity (low flows, high flows) problems	Economic growth Population growth Climate change Water policy implementation	Navigation Agriculture Industry Drinking water

INTEGRATOR 3 – Development of conceptual models for decision makers and the integration of AquaTerra results

The model of the interaction between socio-economic activities and water-soil physical modelling was further developed in the case studies selected in INTEGRATOR 2 (see above):

- The Geer Catchment (Meuse, Walloon region) – diffuse nitrate pollution in groundwater;
- The Kempen Area (Meuse, Flanders region and Netherlands) – heavy metal pollution in groundwater; and
- The central Ebro (Spain) – water resources and salinity.

In these case studies, the cost effectiveness and benefits of management options proposed to mitigate key environmental issues were assessed. For each case study, linkages with the results of other AquaTerra Sub Projects (i.e., BASIN, TREND, HYDRO, COMPUTE and FLUX) were established (cf. the example of the Geer basin, shown in Figure 9.2). The development of the case studies demonstrated how it was possible to integrate the multidisciplinary results produced in AquaTerra and enabled recommendations for further model development such as the need to define a common objective or issue, and the need to develop common understanding among multi-disciplinary teams and to involve stakeholders closely from an early stage. The case studies illustrate how a combined approach of biophysical analysis with economic analysis can serve as a sound basis for policy and river basin management decision making, as it describes the system and predicts changes within it using an holistic environmental/socio-economic approach.

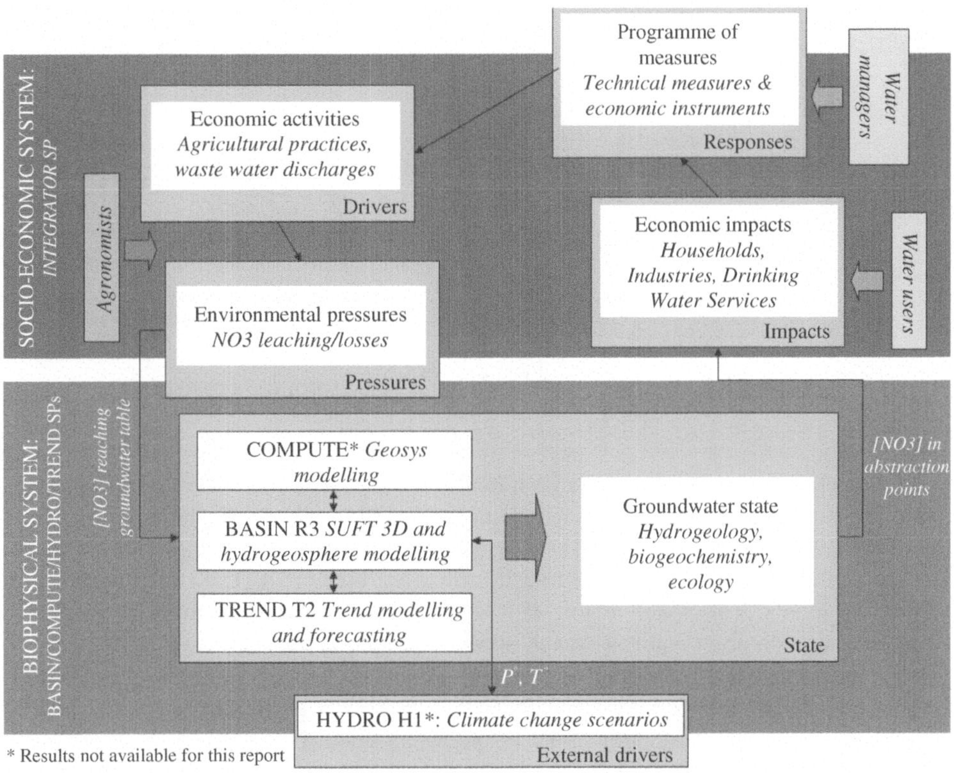

Figure 9.2 DPSIR scheme relationship between INTEGRATOR, other AquaTerra sub-projects and stakeholder on the Geer case study.

A three-stage integration methodology was developed in order to deliver the main results of AquaTerra to selected end-users, independently of the models developed within the case studies (project deliverable I3.12):

(1) Development of a scientific interface to identify key findings of AquaTerra;
(2) Development of an end-users interface to determine key stakeholders' demand;
(3) Correlation of AquaTerra key findings and stakeholders' demands (see Figure 9.3).

The key findings of AquaTerra, validated by two rounds of consultation with WP leaders, are summarized in 53 fact sheets pertaining to the data and knowledge generated, and 43 factsheets related to tools produced as part of the overall project. Stakeholders workshops organised by INTEGRATOR, BASIN and EUPOL and the results obtained were used to compile the stakeholders' demands related to the overall issue of River Basin Management (RBM), plus the stakeholders' specific requests to include the 7 cross-cutting issues most relevant for RBM (scale, climate change, changes other than climatic change, the need for a systemic approach, communication and participation, international and inter-basin coordination and collaboration and the implementation of the WFD). Each of the stakeholder's questions were then directly connected to the most relevant AquaTerra key finding fact sheet(s). Based on AquaTerra results, the key parameters involved in each issue are identified, and recommendations as to how to improve the way the issue can be addressed are given.

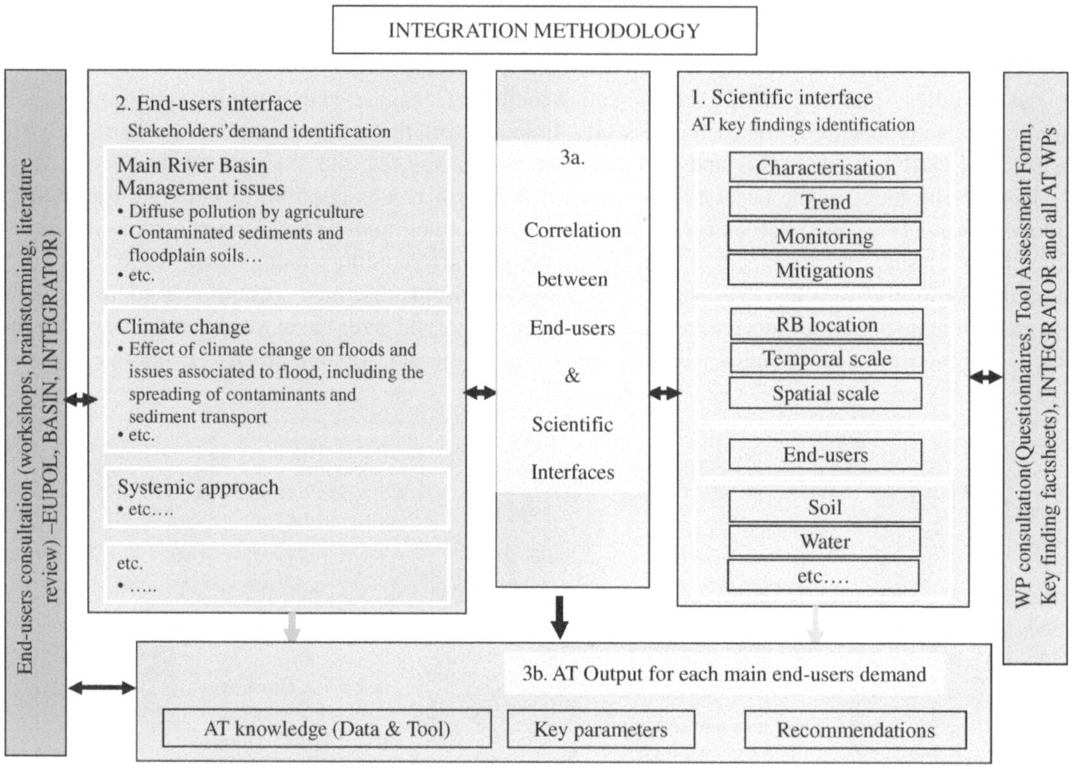

Figure 9.3 Integration methodology of AquaTerra (AT) results.

In order to present AquaTerra science in a manner that makes it of use and interest to as wide a range of potential end users as possible, to place AquaTerra science in an environment where it can be easily and quickly accessed by as wide a range of end users as possible and to place AquaTerra science in an environment where it can be quickly and easily linked to other similar science, the AquaTerra On-line Information System (ATOIS) web portal was developed (see Figure 9.4, and for more details project deliverable I3.13). It consists of an outward-looking portal, presenting the information of AquaTerra in a digested way for stakeholders. The core elements are:

(i) An AquaTerra encyclopaedia, consisting of synthesis reports (technical summaries) of the information from different AquaTerra deliverables organised under key themes such as soil, groundwater and contaminants.
(ii) Online catalogues containing information about resources produced directly as a result of AquaTerra, as well as further relevant information from beyond the project that would be of value to the target end-user group.
(iii) All AquaTerra factsheets (knowledge and data factsheets and tool factsheets).
(iv) Direct access to homepages of other relevant scientific projects and resources.

The final aspect of the work of INTEGRATOR was the identification and assessment of the inter-relationships between AquaTerra and other scientific (project deliverable I3.5). Interviews with AquaTerra work package leaders, along with internet searches and conference workshops identified potential and existing collaborations. AquaTerra developed a large and successful network of knowledge and contacts at the European, national and local scale. Strong collaboration was further developed with projects including RISKBASE, SEDNET, MODELKEY, AquaTrain, NORMAN, SWIFT and EUGRIS and opportunities for collaboration with SPI-WATER, WISE-RTD and NEWATER were also explored.

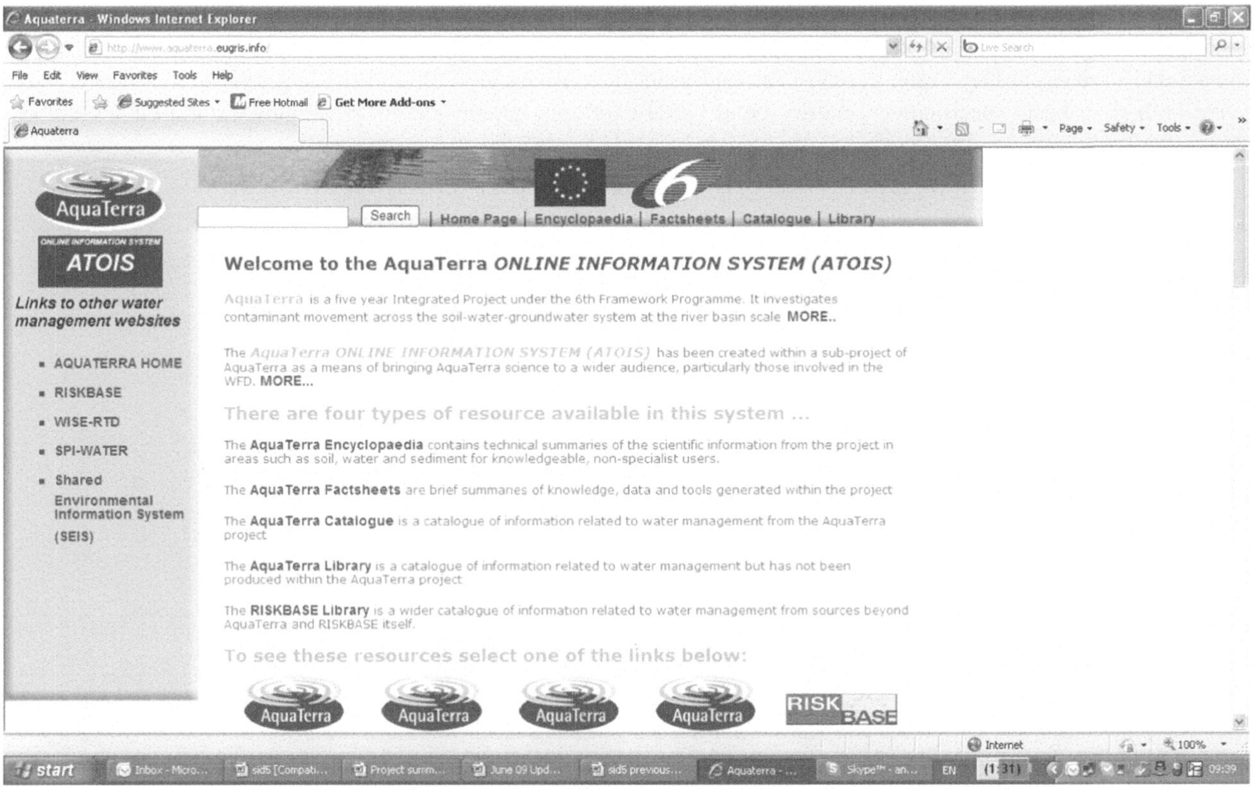

Figure 9.4 Home page of the ATOIS website.

IMPLICATIONS AND APPLICATION OF RESULTS

Many INTEGRATOR achievements have been applied in the wider context of AquaTerra:

- The development of the case studies (in the Kempen, Geer and Ebro catchments) demonstrated the application of the DPSIR conceptual models. These conceptual models integrate environmental, social and economic data and provide a useful tool for decision-making in river basin management.
- The implementation scales of the conceptual model developed within INTEGRATOR include Europe (Climate & Policy drivers), the river basin (socio-economic drivers) and the local/ catchment area. This is shown in Figure 9.5.
- The WP leaders' consultation undertaken by INTEGRATOR as part of the development of the AquaTerra factsheets enabled the identification and compilation of key findings within different sub-projects. It also helped to assess their relevance with respect to addressing river management issues and identified end-users who could benefit from AquaTerra key findings. The development of the ATOIS portal and the key finding fact sheets have brought the results to the attention of a wider audience of stakeholders. They shall be useful tools to support decision-making among river basin management stakeholders.
- The principles and structure of workshops to communicate such information to stakeholders was demonstrated through the workshops undertaken in the four selected river basins as part of the INTEGRATOR SP. The experience of conducting such workshops is of particular interest and relevance to those involved in implementing the WFD, as stakeholder involvement is an obligation within this Directive.

The development of DPSIR within AquaTerra was conducted using a top-down approach. The conceptual model was first created at the river basin level and then used and refined at the local level. Conversely, a bottom-up approach was adopted to identify and locate problems related to the soil-water system and then to produce policy recommendations at a European level. The implementation of DPSIR at the local and regional level was built through close involvement of local/regional actors and stakeholders through workshops and meetings.

The economic method applied in the case studies may be generalised to river basin level. However, this needs to be adapted according to the type of environmental issues being discussed (e.g., point source and diffuse pollution will not be analysed with the same economic models) the type of main contributing pressures involved and specific conditions within the river basin of concern.

The conceptual models developed within INTEGRATOR can be applied widely in different river basins. The methodology to do so is clearly described in Chapman *et al.* (2007). The INTEGRATOR case studies serve as illustrations of the approach and also provide practical policy recommendations. The collation and combination of the varied output of results through INTEGRATOR 3 is a means by which this information can be brought to a wider audience and used according to the needs of the end-users themselves. ATOIS and the key finding fact sheets (which provide both river basin-specific and generic sound based data) can be considered as valuable tools in this context.

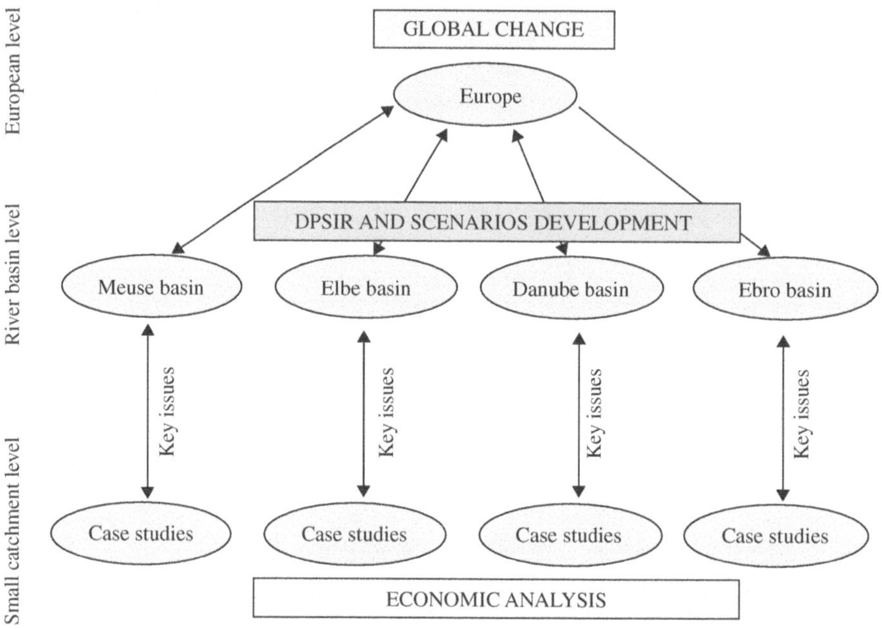

Figure 9.5 Spatial scales in INTEGRATOR sub-project.

Several methods developed through the course of the work packages, in particular in INTEGRATOR 3 sub project, could be applied in other European projects. Especially the approaches used to combine diverse results or to identify potential collaborations could be widely applied.

Drivers identified in the DPSIR model such as climate and policy changes are of global concern. DPSIR is one of many methods by which such global issues can be related to smaller scale regional, national and local issues. As such it can be applied in other areas of the world beyond Europe.

CONTRIBUTIONS AND POLICY RECOMMENDATIONS FOR RIVER BASIN MANAGEMENT

The experience gained from the process of integrating AquaTerra results that was undertaken in INTEGRATOR 3 provides recommendations with respect to each key finding and also with respect to each key question raised by potential users (see the factsheets of project deliverable I3.12). Some of these recommendations were relevant to river basin management or policy, in particular in relation to the fields of data acquisition, monitoring measures and data management, existing water policies and the need for new policies.

Moreover, the development of the DPSIR integrated conceptual model and associated coupled socio-economic and biophysical modelling has allowed recommendations for river basin management to be proposed.

These recommendations can be applied to the local, national, European or global level and a selection of opportunities to do so is presented below.

Locally/nationally

- It is important to clearly assess the expected changes in the economy, climate and policy arenas before deciding on management options. Flexibility in management options with regard to the uncertainty of future events is highly important.
- Communication between information providers and stakeholders is vitally important so as to understand local and regional problems clearly and to advertise the range of methods and approaches available to help solve them.
- Research and activities should also have a local aspect, or at least involve and inform local stakeholders and develop research work based on applied case studies.

- Make sure that the project team works effectively together, particularly in instances where a project requires a multi-disciplinary team combining scientists of different backgrounds and potentially economists, social scientists and policy makers depending on the nature of the work.
- The development of integrated applied or conceptual models makes sense only if they address clear objectives and issues to be set out from the start with all parties involved in natural resources management.
- In order to adequately simulate natural resources and to have a holistic view of the characterisation of the system and prediction of its behaviour, it is necessary to consider both anthropogenic and environmental parameters. As a consequence coupled socio-economic and biophysical modelling is a very powerful decision making support tool.
- The integration of AquaTerra results integration provided further recommendations on data acquisition, monitoring and measurement techniques and data management.

In a European context

- In terms of the development of effective links between the soil, sediment and water components of river basins, many policies deal with water status as a result of the WFD. However, fewer policies address the soil component. The proposed Soil Framework Directive is one policy instrument by which this disparity could be addressed, although the Directive has not (as yet) been universally accepted by the EU member states for a variety of reasons: this makes it all the more important that integrated work linking soil and water research is encouraged.
- Improve knowledge of soil and sediment processes and harmonize data between countries at river basin level. Within AquaTerra and other monitoring programs the characterisation of the soil-sediment-water system at a river basin level was supported by numerous data. However, in some cases these data were heterogeneous between different providers (e.g., in some cases there was a wide range of accuracy in the data provided to AquaTerra scientists due to the different methods applied). In general, data and information related to the water component of the system are widely available (due the work undertaken as a result of the implementation of the WFD). Most data are displayed at the river basin or at the sub-basin scale, particularly for water quality. However, data and information about soils and sediments is mostly available at country level or for a very specific area: data on soil degradation and sediment are sparse, very heterogeneous and not always comparable (e.g., the methods used to measure sediment quality differ from one country to another). Economic data are available at a range of different levels (European, national or regional). It would be a challenging but important task to unify these data from different sources and make them centrally available for future decision making and modelling.
- Water policies defined at European level (i.e., the WFD) tend to promote analysis of the system at the river basin scale. However, there are limitations to the river basin scale approach, as in some cases, important biophysical processes are mainly relevant for regional or sub-regional water bodies.
- AquaTerra results have been used to provide feedback on existing policy implementation, such as the incapacity for some water bodies to meet the deadlines for water body characterization or good ecological status set by the WFD due to the temporal consequences of slow hydrogeological systems necessitating a long characterization time or a long response time for the effects of mitigation measures to take place.
- AquaTerra research identified fields where new regulations should be implemented, namely the sources of diffuse pollution; water resource management (in particular water scarcity); the reduction of discharges of pharmaceutical compounds to surface water; qualitative and quantitative characterisation of sediments.
- There is a necessity to develop a common policy throughout Europe on hydrological and environmental data distribution and to standardize the format of the dataset. Since AquaTerra began this process has actually begun and is currently being undertaken through measures such as the Shared Environmental Information System (SEIS) and the Water Information System for Europe (WISE).

OPEN & NEW QUESTIONS FROM INTEGRATOR

Several new questions have emerged from the INTEGRATOR research:

- How can various management scales be conciliated when dealing with several components of the soil-sediment-water system and their interaction? During the INTEGRATOR workshops, there was a discussion about the level at which sediment-soil-water issues should be addressed, i.e., local, regional, national or international. An important issue at the regional / national level is not necessarily viewed as an important issue at the river basin scale. The response and management concerns may also vary between the different levels, especially in the case of trans-boundary river basins.

- How is it possible to take into account various points of (non scientific) view and integrate them into scientific work?
- How can river basin management not only respond to environmental disturbances, but build up an early warning and prediction system that drives future planning?

Political aspects are very important in identifying environmental issues. Stakeholders affected or concerned by environmental problems may have a perception that differs from the view of a scientifically based approach. Science often provides detailed information to support decision-making, though this is not necessarily taken up effectively through the policy making process. Political decisions may also be influenced by societal pressure, which may or may not be informed by scientific knowledge.

The tools and methodologies (methodological framework, workshop organization methodology, AquaTerra key finding fact sheets) developed within AquaTerra can be applied in contexts outside the project (i.e., for key management issues and river basins that have not been studied in the context of AquaTerra). AquaTerra results can also be further applied by promoting dissemination of information through ATOIS, scientific and non-scientific literature and workshops.

REFERENCE

Chapman, A., Brils, J., Ansink, E., Hérivaux, C. and Strosser, P. (2007) Conceptual models in river basin management. In: *Groundwater Science and Policy: An International Overview*. Philippe Quevauviller (ed.), Cambridge: Royal Society of Chemistry.

Chapter 10

EUPOL – Linking policy demands and research

A. F. L. Slob, M. Rijnveld, A. S. Chapman and P. Strosser

KEY OBJECTIVES

The EUPOL subproject of AquaTerra aimed to link policy demands to the scientific information on processes in river basins at a variety of scales being generated by AquaTerra in order to bring the scientific output to a wider, more influential field. Although a small subproject within AquaTerra, EUPOL was the main element working on this aspect of the project. The Description of work described the main objectives of EUPOL as follows:

To contribute to a future EU policy framework covering Climate Change, Land Use and water, groundwater, sediment and soil management, by supporting the development and elaboration of a consistent policy framework and facilitating the interaction between AquaTerra scientists and policy makers. This is done with the aim of:

- Identifying the impact of the AquaTerra results on the Water Framework Directive, the Groundwater Daughter Directive and the proposed Soil Framework Directive.
- Providing advice for future development of integrated soil and water quality management and policies in Europe based on this analysis.

KEY ACHIEVEMENTS

- Identification of policy questions.
- Scientific framework for linking questions to deliverables (impact of research).
- Recommendations towards integrated river basin policy.
- Facilitating conversation between policy makers and researchers.

Knowledge production and use is regarded by EUPOL as a social process. It is argued that following this point of view and regarding the level of the river basin scale, the research approach should be more holistic and socially accountable. Consequently, it should emphasize the interaction between the different communities involved (scientists, stakeholders, researchers). To begin this process it was necessary to identify the nature of the interaction through the key topics of concern to policy makers. Therefore, in the EUPOL subproject a scientific framework was developed, to first identify the major questions of policy makers and then to assess how such questions could be, at least in part, addressed by AquaTerra research.

A range of 54 policy questions were identified from four brainstorm sessions, questionnaires and interviews (project deliverable E1.1c, scientific framework, see www.eu-aquaterra.de). These were grouped into broad categories of reference, to show which scientific areas were of particular interest to policy makers. The distribution of questions among these categories is shown in Figure 10.1. The questions were mostly related to land use, river basin management, water and climate change, while questions related to agriculture and soil were the least common. However, European soil policy is less developed than water policy at present and climate change and river basin management are currently important topics of discussion: in this context the distribution of policy questions reflected the level of interest in these areas reasonably accurately.

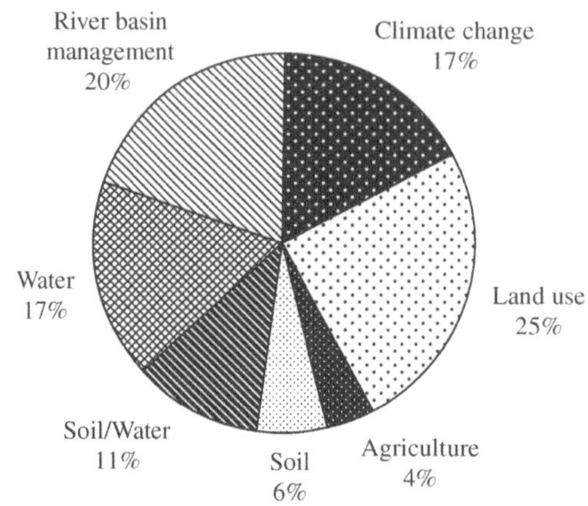

Figure 10.1 The distribution of policy questions by subject (A.F.L. Slob, M. Rijnveld, A.S. Chapman, P. Strosser, Challenges of linking scientific knowledge to river basin management policy: AquaTerra as a case study, Environmental Pollution, Volume 148, Issue 3, August 2007, Pages 867–874).

Having identified the nature of policy demand through these questions, the next stage of the EUPOL project was to assess how the supply of scientific information from AquaTerra could be linked to it. The Scientific Framework that was developed to link them is effectively a matrix that moves from the intellectual starting point of each subproject, through its individual research deliverables to the policy needs of stakeholders and policy makers and vice versa (Chapman *et al.*, 2009). The subprojects were all assessed as to how well their work may answer the policy questions with their respective deliverables. In total there were 286 research deliverables listed for the relevant subprojects over the first 30 months of AquaTerra, representing its scientific output; each one was ranked as being of high, medium or low relevance to each policy question. The results of this work, categorised by workpackage, are shown in Figure 10.2 below.

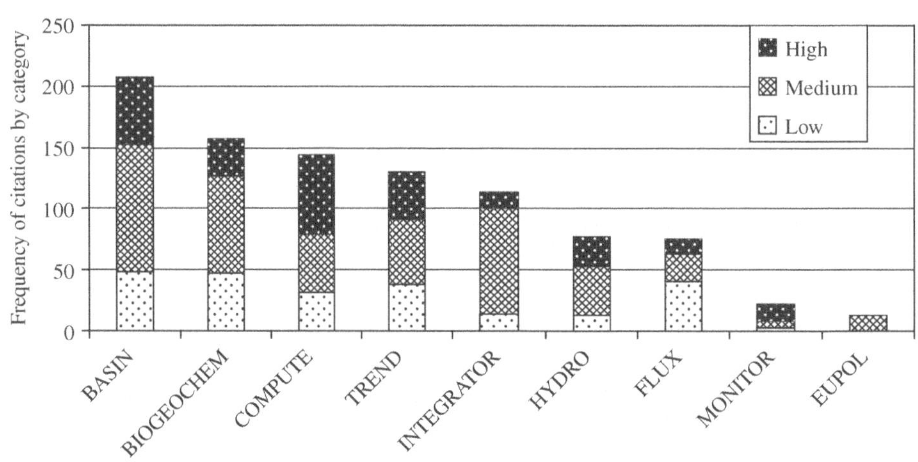

Figure 10.2 The relevance to the policy questions of the deliverables in each work package (E1.1c).

According to the results from the Scientific Framework several questions could be answered very well by combining information from research deliverables, although a match in itself is no guarantee for knowledge uptake by policy makers. Much of the information generated by AquaTerra must subsequently be interpreted, digested, restructured and combined with information from other sources before it can be considered of genuine use and value to policy makers who have to use it in order to make an informed decision on, e.g., river basin management measures.

Soil degradation, Water Quality, Land Use and Climate Change are policy issues of contemporary urgency, which are classically 'wicked' and with various strategic responses. Both the science and policy communities have individual responsibilities for problem diagnosis, knowledge generation, intervention planning, and evaluation of impacts. From a system perspective, the responsibility for the environment is scattered among various agencies and policy areas.

Yet, they have a communal responsibility to ensure effective communication and interaction between the various constituencies and to develop appropriate approaches to integrate systemic knowledge into effective policies. When taking a closer look at how environmental policies are arranged in European countries, we see that it does not reflect the unity of the natural river basin system. While the river shows a dynamic interaction between components such as groundwater, sediment, soil and surface water, our societies have organized the stewardship of land and rivers in more or less static parallel policy domains and science disciplines. In general, the use of (scientific) knowledge in policy processes is often not very impressive and factors like presentation, timing, (dis-) similarities with policy-makers' beliefs are important for the utilisation of information. The establishment of connections between both communities is especially problematic for complex societal issues, such as river basin management, as the policy processes have to deal with many participants with a different framing of the problem.

We have recommended that developing a shared view of the river basin contributes to the creation of a social context. That will facilitate the convergence of research results and provide policy makers with a language to interact with the research community. The Water Framework Directive provides an opportunity to develop a common ground among stakeholders of what the main issues in the river basin are.

IMPLICATIONS AND APPLICATION OF RESULTS

The existence of knowledge stored in research deliverables, even when it could potentially provide answers to questions from policy makers, is in itself no guarantee for knowledge uptake by policy makers. A key barrier in the uptake of scientific results in decision making lies in the fact that different stakeholders have a different understanding of the river basin. E.g. fishermen, farmers, city authorities, water boards, provinces, regions, biologists, geologists, mathematicians and economists all have a very different perspective on what the river basin is. Their varying views on reality influence how they see the river, how they define the natural resources, how they perceive risks and what knowledge is valuable or not. They simply do not have a common understanding of what the river basin is and consequently lack a common basis for sharing information.

We have suggested that developing a shared understanding among stakeholders about what the river basin is, contributes to the creation of a social network with researchers and policy makers. A shared idea of what the river is could be constructed of basic elements such as

- natural resources (water, sediments, soil),
- ecosystem services (storage, decomposition, recreation, drinking water),
- causal relations in the river basin system (trends, interventions), and
- present and future risks (vulnerability, resilience).

Although different stakeholders may still have their own perception of these basic elements, they can serve as boundary objects between different groups of stakeholders. Star *et al.* (1989) define boundary objects as *"tangible artefacts or object-like forms of communication that inhabit several intersecting social worlds and satisfy the information requirements of each of them. [...] They have different meanings in different social worlds but their structure is common enough to more than one world to make them recognizable, a means of translation"*. Although newly constructed basic elements describing the river basin would help to acquire a mutual understanding, "the tacitness of much knowledge often makes codification, transfer, and subsequent replication of routines and standard operating procedures difficult" (Bechky, 2003). Consequently, Bechky advocates to speak of "knowledge transformation processes", eliciting that 'something happens' with knowledge that crosses boundaries, being re-shaped and integrated to practices of different communities. Nevertheless, a successful approach to knowledge transformation starts with the creation of "common ground between the communities involved". A process in which stakeholders develop a shared description of the basic elements of a river basin should facilitate the development of a 'common ground'.

An important condition for creating common ground is the existence of a social context. A conclusion that is commonly cited in literature is that an intensive interaction between researchers, policy makers and stakeholders from an early stage appears to improve the uptake of scientific information (Lavis *et al.*, 2002; Lomas *et al.*, 2003; Joyce, 2003). Landry *et al.* (2001) found that the extent of the dissemination efforts undertaken by researchers is a good predictor for the uptake of scientific information by policy makers and that the context of the researchers as well as that of the policymakers influences knowledge utilization. They have concluded that knowledge utilization is much more dependent on factors related to the behavior of researchers and the users' context, than on the attributes of the research products. Therefore, it is necessary to look at science policy interfaces as social processes (Slob *et al.*, 2007; van den Hove, 2007). Interaction between policy makers, researchers and stakeholders is a key element and should be organized in such a way that questions and answers can be generated simultaneously.

Several lessons can be drawn for the science-policy interface that are relevant for a new policy direction in river basin management:

- Interaction between scientists of different disciplines is needed. This asks for better integration of disciplinary results and trans- and multidisciplinary research methods (Totslandal *et al.*, 2007; Slob *et al.*, 2007).
- Interaction between scientists and users of scientific results is important for the development of effective and supported policy (Totslandal *et al.*, 2007; Slob *et al.*, 2007; Slob and Ellen, 2007; Quevauviller *et al.*, 2005)
- Stakeholders should be involved from an early stage in the scientific production process as well as in the policy process (Slob *et al.*, 2007; Slob and Ellen, 2007; Totslandal *et al.*, 2007)
- This interaction should be shaped as a learning process that allows to develop the understanding of the system and to fine-tune the measures that should be undertaken. Monitoring of effects of these measures is needed to better understand the impacts of human interventions in the system (Bouma and Droogers, 2007; Slob *et al.*, 2007; Slob and Ellen, 2007).

Following these guidelines, intensive interaction with stakeholders, policy makers and scientists should be organized, to support the uptake of scientific results by policymakers and stakeholders. If policy-science linkages are designed using a similar format across different disciplines, this will aid consistency and understanding at an inter-departmental level within major political institutions such as the EU. As a result, the Scientific Framework, or a development of it has the potential to be a very valuable means of improving understanding and increasing the transparency of local, national and international policy and decision making.

CONTRIBUTIONS AND POLICY RECOMMENDATIONS FOR RIVER BASIN MANAGEMENT

The EUPOL team has developed recommendations for AquaTerra as a research project to better contribute to future EU policies:

(1) Crossing boundaries between scientific disciplines involved in multidisciplinary research projects like AquaTerra, will not organize itself spontaneously, but has to be organized. Sharing the responsibility for collaborative production of (policy oriented) knowledge and its application is not a given. There are too many reasons for professionals on either side of the boundary, whether between scientific disciplines or between science and policy, to remain in their quarters and avoid (direct) interaction with each other. We recommend therefore that in such big, multidisciplinary research projects like AquaTerra, a specific dedicated work package should have the task to foster connectivity between scientific disciplines and to organize the process of interaction between these disciplines in a meaningful way for joint conceptual modelling and conclusions.

(2) Involving the potential users (policy makers, politicians, companies, etc.) from the startis needed for the Information generated by AquaTerra to be interpreted, digested, restructured and combined with information from other sources before it can be considered of genuine use and value to others.

(3) To develop a shared understanding among participants about the research subject, intensive interaction in workshops with stakeholders, policy makers and scientists should be organized. This interaction should be shaped as a learning process that allows to develop a common understanding of the river basin system and to fine-tune the measures that should be undertaken. Monitoring of effects of these measures is needed to develop a common understanding of the impacts of human interventions.

(4) The iterative, cyclical basis of the Water Framework Directive provides an opportunity to develop a common ground among researchers and various stakeholders during the second round of River Basin Management Plans, which will be initiated before the 1st management cycle ends in 2015. Active participation will have to be organized around various issues. It will be helpful to share perceptions on water, sediments, soil, storage, decomposition, recreation, vulnerability, etc. It allows all kinds of stakeholders to learn from both past successes and failures, as well as to learn from experience and knowledge of others. It could be approached as a joint development programme in which knowledge about the river basin is being co-created by scientists and practitioners.

OPEN & NEW QUESTIONS FROM EUPOL

Over the years we have developed an understanding of the mechanisms behind the use of research results in daily river management practices. Essentially, shaping the boundaries between academic fields and that of academics and end-users seems to be crucial for a successful exchange of knowledge. We assume that various actions within AquaTerra may have been very successful means of creating common understanding in a social context (such as the Danube Sampling cruise or the work done in BASIN). Yet, systematic research of successful examples has not been done as it was beyond the scope of EUPOL. Therefore, from the perspective of improving the uptake of research results in policy making, the main questions that remain after 5 years of AquaTerra research are:

(a) To what extent have research results been used in European river basin management policies?;
(b) What examples have proven to be successful?, and;
(c) Which factors have determined the use of knowledge?

These questions could make part of an evaluation of the societal relevance of AquaTerra. The answers could provide insight in the mechanisms behind the uptake of research results in policy making and management.

REFERENCES

Bechky, B.A. (2003) Sharing meaning across occupational communities: the transformation of understanding on a production floor. *Organization Science* **14**(3), May-June, 312–330.

Bouma, J. and Droogers, P.P. (2007). Translating soil science into environmental policy: a case study on implementing the EU soil protection strategy in the Netherlands. *Environ. Sci. Policy*, doi:10.1016/j.envsci.2007.02.004.

Chapman, A.T., Slob, A., Rijnveld, M. and Merly, C. (2009) Evolution of methods to link science and policy: The experience of EUPOL. *Water System Science and Policy Interfacing*, P. Quevauviller, (ed.), RSC Publishers.

Joyce, L.A. (2003) Improving the flow of scientific information across the interface of forest science and policy, *Forest Policy and Economics* **5**, 339–347.

Landry, R., Amara, N. and Lamari, M. (2001) Utilization of social science research knowledge in Canada. *Research Policy* **30**, 333–349.

Lavis, J., Ross, S., Hurley, J., Hohenadel, J., Stoddart, G., Woodward, C. and Abelson, J. (2002) Examining the role of health services research in public policymaking. *The Milbank Quarterly* **80**, 125–154.

Lomas, J., Fulop, N., Gagnon, D. and Allen, P. (2003) On being a good listener: setting priorities for applied health services research, *The Milbank Quarterly* **81**, 363–388.

Quevauviller, P., Balabnis, P., Fragakis, C., Weydert, M., Oliver, M., Kaschl, A., Arnold, G., Kroll, A., Galbiati, L., Zaldivar, J.M. and Bidoglio, G. (2005) Science-policy integration needs in support of the implementation of the EU water framework directive. *Environmental Science & Policy* **8**, 203–211.

Slob, A. and Ellen, G.J. (2007) The changing role of science in river basin management. In: Remediation of contaminated sediments-2007, *Proceedings of the Fourth International Conference on Remediation of Contaminated Sediments*, Battelle Press.

Slob, A.F.L., Rijnveld, M., Chapman, A.S. and Strosser, P. (2007) Challenges of linking scientific knowledge to river basin management policy: AquaTerra as a case study. *Eviron. Pollut.* 867–874. doi:10.1016/j.envpol.2007.01.048.

Star, S.L. and Griesemer J.R. (1989) Institutional ecology, translations and boundary objects: amateurs and professionals in Berkeley's museum of vertebrate zoology, 1907–39". *Social Studies of Science* **19**(4), 387–420.

Totlandsdal, A., Fudge, N., Sanderson, E., van Bree, L. and Brunekreef, B. (2007) Strenthening the science-policy interface: experiences from a European Thematic on Air Pollution and Health (AIRNET). *Environmental Science & Policy* **10**, 260–266.

van den Hove S. (2007) A rationale for science-policy interfaces, *Futures* **39**(7), 807–826.

Chapter 11

AquaTerra's achievements and recommendations to river basin management in brief

M. Finkel, J. Barth and P. Grathwohl

A good start for looking at achievements in AquaTerra is to consider water as the phase that links between all compartments studied (atmosphere, soils, aquifers, surface runoff and sediments). It is the transport and reaction medium in almost all environmental processes and was considered in an integrative way in AquaTerra. Flow of water was modelled accounting for changes in precipitation and runoff to be expected in different climate projections to assess the influence on the water cycle.

For instance, the HYDRO subproject demonstrated that any single model simulation of future climate may represent only one of many different future climate outcomes. Therefore *several different models* need to be compared to historical climate trends to assess their possible ability in projecting future climate scenarios. Climate projections for 2070 by the HYDRO team indicate more frequent summer droughts of long duration (>6 months) in Southern Europe and more short extreme precipitation events across the whole of Europe, while longer extreme precipitation periods will likely increase in frequency in Northern Europe.

River basin management should concentrate on the reduction of current vulnerabilities to extreme events; i.e., increasing the robustness of our river basins and water supply systems to flood and drought events while bearing in mind that these types of events are likely to increase in frequency and magnitude in the future.

Such drastic climate alterations have several implications for other findings in AquaTerra: For instance, these alterations will significantly influence the transport of pollutants in river basins. That is, more mobility of pollutants and mass transfer between compartments can be expected in areas prone to higher precipitation in northern Europe. In BIOGEOCHEM (BGC) we already found that shifts from oxic to anoxic conditions after floods cause release of organic colloids and in turn persistent organic pollutants and redox sensitive inorganic compounds are mobilized. Such effects will become even more pronounced in areas of more frequent flooding as colloid-facilitated transport affects the speciation, distribution, and bioavailability of contaminants within the floodplain ecosystem. After long droughts preferential flow will enhance vertical transport of pollutants to groundwater.

River basin management therefore needs to highlight floodplains when designating areas for re-naturation and flood control storage.

The second implication is that such climate changes will affect deposition of airborne pollutants across Europe. Findings by FLUX and MONITOR showed that persistent organic pollutants are deposited in largely different patterns across Europe. The lowest deposition rates of polyaromatic hydrocarbons (PAHs) as a proxy for atmospheric deposition were found in the Ebro- and the highest in the Danube Basin. One of the key drivers for these different patterns may be drastically differing precipitation patterns between these basins. If so, areas with more precipitation will probably receive higher deposition rates of persistent organic pollutants in the future. On the other hand, we may find more mobilisation of particle- and colloid-associated pollutants in areas affected by floods. These effects of variable deposition and mobilisation need more attention in the future.

In addition to Europe-wide and global hydrologic models, high-resolution local monitoring of a catchments provided better understanding of the functioning of small but still complex groundwater systems. For instance, the Brévilles spring catchment near Paris was established as one of the best-instrumented agricultural catchments in AquaTerra and is exemplary for European groundwater research with respect to key compounds such as pesticides. Its investigation has provided new, detailed knowledge about consequences of pesticide applications and showed that even polar organic pollutants such as Atrazine and its metabolites can remain in the subsurface for long periods of time up to many decades if not centuries. Scaling up to river basins would likely multiply residence times. On the other hand, longer time frames may offer chances for degradation and metabolite formation. Strontium isotopes coupled with Mg/Sr and Ca/Sr ratios in groundwater in this catchment successfully demonstrated the ability to trace agricultural impact such as input of fertilizers. The latter significantly modify the natural Sr isotopic signatures and helped to delineate different groundwater bodies. This new isotope techniques allow tracing water and solute pathways and their recharge patterns and helps to provide scenarios for the hydrological functioning of underground systems.

The use of isotope analysis as part of environmental monitoring is one of the recommendations for use of modern environmental tools because it can distinguish anthropogenic and natural sources of contamination.

Microbiological work revealed functional activities and composition of microbial communities in various locations in the Brévilles, Danube, Ebro and Elbe basins. Detailed investigations of Ebro river sediments showed that bacterial communities change seasonally and spatially. This implies that turnover of pollutants varies locally, depending on temperature, pH and other key environmental parameters. As one target compound, the microbiology team of AquaTerra studied the bioavailability and degradability of nonylphenol in sediments and related it to the estrogenic toxicity of the sediment. A key finding in this context was that microorganisms are able to degrade desorbed nonylphenol under laminar flow conditions.

The above summary outlines that it is important to consider fluxes and long-term fate of persistent organic pollutants in an integrated manner as undertaken by the subprojects TREND, MONITOR and FLUX. To satisfy this demand, a new database for evaluation and interpretation of net exchanges and fluxes between environmental compartments has been started in AquaTerra. It contains pollutant concentrations from passive air samplers, atmospheric deposition, water and soil samples in the catchments of Danube, Elbe, Ebro and Brévilles. Based on these novel data, diffusive soil-air fluxes were calculated for polychlorinated byphenyls (PCBs) in the Ebro, Danube and Elbe Basins to establish an important end member of the pollutant chain studied in the project and further on. This research revealed net deposition to soils of key persistent organic pollutants such as PCBs. Volatile compounds, however, revolatilize from soils. The work also showed that degradation is the most important loss process for PCBs next to runoff to freshwater. Other findings on persistent organic pollutants (POP) in AquaTerra indicate that more than 90% of the incoming PAHs remain in the catchments and mostly accumulate in the top-soils. Strong sorption of such compounds to combustion-derived carbonaceous material was found to be the cause for limited bioavailability and also their strong persistence. This shows that combustion – a major source of pollution from power plants, traffic and household heating – simultaneously influences pollutant behaviour in the environment by uptake POPs on black carbon or soot. This is strongly linked to transport by particles that ties back to important mechanisms of POP transport through colloids in floodplains.

As opposed to the findings about diffuse pollution, several regional hotspots of pollution were also monitored within AquaTerra, for instance the Flémalle site in the Meuse Basin or the Schachtgraben in the Bitterfeld region in the Elbe Basin. In the latter, the surface- and groundwater interaction was characterised. Concentrations of for instance chlorobenzenes and related compounds were found to be significantly lower in groundwater and river water when compared to the pore water of the streambed.

The rivers' streambed may constitute an important contaminant and turnover reservoir in environmental pollutant dynamics. The status of these reservoirs therefore deserves more attention in river basin management and investigation.

The observed concentration differences between groundwater and the hyporheic zone confirm that this intermediate compartment is a key for storage and transformations. As a general finding for hotspot areas, it was confirmed by research of the BIOGEOCHEM team that carbonaceous materials (e.g., black carbon) act as strong sorbents for organic pollutants. Amendment of sediments with activated carbon is currently under discussion as a cost-effective remediation strategy. Nonetheless, this needs to be investigated with care as the immobilisation needs to be stable for long time periods of centuries and it needs to be understood that such measures compete with biodegradation and thus stabilize pollution.

General recommendations by FLUX research are that existing monitoring networks need to be adapted to account for state-of-the-art geochemical tracers and isotope methods. Rivers are integral sources of information and their variable discharge has to be taken into account for pollutant and material flux considerations. Efficient response sampling should also be increased in the future as flood events have a significant but so far poorly characterised transport function. They should therefore be included in monitoring programs. Floods are also responsible for the largest part of suspended sediment transport and no clear information exists on the magnitude of such transport for considerations of erosion and pollutant mobility. This means that suspended sediments have to be also monitored and this should be accounted for in regulations such as the EU Water Framework Directive. The best cumulative information for basin-wide transport is available at river mouths that should be routinely monitored for suspended sediment (and attached pollutants) as well as dissolved load fluxes.

While the above diffuse and hotspot pollution scenarios were mostly carried out on proxy compounds such as PAHs and PCBs, the subproject MONITOR developed new methods for pesticide, pharmaceuticals and flame retardant analysis. These enabled efficient quantification of these compounds and their derivatives. The analytical protocols are for instance based on solid phase extraction (SPE) and liquid chromatography – tandem mass spectrometry (LC-MS-MS) and provide reliable measurements at trace levels. This is highly relevant for diffuse pollution where low concentrations and trends in time and space need to be reliably detected. For further improved sampling in the field, a bulk wet and dry deposition passive sampling device was used. As a breakthrough in passive sampling, two types of passive samplers were also successfully validated for measuring polybrominated diphenylethers (PBDEs), a compound group that is widely used as flame-retardants. For metals, magnetic proxy screening outlined local variations of relative pollution loads within the Danube and Elbe Basins as demonstration cases. The resulting 3D maps of magnetic parameters serve as proxies for industrial pollution and allow improved target selection of sampling and testing locations. Neural network methods were also used to analyze the data obtained within this context with the aim to optimize sampling strategies. Such calculations allow estimation of the number of locations and time intervals between measurements and other parameters for optimizing sampling strategies in environmental monitoring.

To obtain sufficiently comprehensive pictures of the chemical status of a river basin, systematic monitoring of contaminants including selected emerging contaminants and their main transformation products as well as advanced measuring systems for different stages of wastewater treatment plants are recommended.

MONITOR has shown that monitoring data of emerging contaminants are often not comparable from one river basin to another (or one country to another) because of different sampling and analysis approaches.

Harmonization of existing monitoring and analytical protocols in Europe and inter-laboratory comparisons are needed.

For spatial and time trend evaluations the TREND subproject found surprising results showing that repeated applications of manure fertilizers for over 40 years have tripled the sorption capacity for heavy metals in agricultural soils. Nonetheless, Dommel sediments in the Meuse Basin were found to exceed quality criteria for cadmium and zinc in areas where smelters influence the environment largely by heavy metals. When considering the key compartment, soil formation requires long time periods of centuries to millennia while erosion, degradation and sealing usually happens on much shorter time scales.

These findings confirm that protection and sustainable soil use are urgently needed to preserve soil buffer and transformation functions.

A key question of metal availability to organisms was assessed by the "Acid Volatile Sulphide and Simultaneous Extracted Metals" (AVS–SEM) method REF. As another bio-monitoring tool, earthworm abundances and their reaction to metals appeared to be controlled mostly by moisture, clay and organic matter contents rather than by metal concentrations in the soil. The sensitivity of these important ecological soil inhabitants was found to be more prone to changes in soils rather than currently encountered levels of pollution. For aqueous biomonitoring, a clear decrease of toxic effects in water and upper layer sediment was observed at the River Elbe in spring related to high water discharge and algal blooms.

It can be concluded that for an appropriate identification and assessment of ecological risks the set of eco-toxicological tests needs to be carefully and adequately selected to cover an acceptable range of organisms and eco-toxicological responses.

The main achievement of the groundwater research team within TREND was the provision of dating tools to detect trends and trend reversal of groundwater quality. These methods and concepts have been introduced in EU Guidances for the WFD and GWD (EU, 2008) and in the Common Implementations Strategy of the WFD through presentations in EU Working Group C.

Data on concentration trends need to be aggregated over the groundwater body in order to obtain useful results and to reduce the overall effects of aquifer and flow heterogeneity. The assessment of the hydrologic response times of the groundwater systems is a key to the understanding and detection of trends.

In the BASIN team a GIS system that has been developed for interpretation of historical and recent monitoring data in the Ebro basin. In this context, the Ebro case study highlights the need of multidisciplinary investigations to cover all potential impacts of human activities to the environment. This includes water regulation, data on dams, agriculture with chemical inputs and irrigation as well as urbanization and industrialization with the related wastes released into the environment. Such measures can help to weigh natural phenomena versus anthropogenic ones and thus improve recommendations for river basin management.

For the Meuse the model BioChem-DSS was specifically designed for the prediction of heavy metal availability in floodplain sediments under varying environmental conditions (e.g., redox potentials). Ecotoxicological effect studies proved that if heavy metals are present in the pore water, they may also appear in the food chain. On the other hand, the AquaTerra Danube Survey (ADS) is an achievement by itself and has resulted in the collection several hundred sediment, suspended matter and water samples from a 1147 km long stretch of the Danube River. It also laid foundations for further large-scale sampling campaigns in this river system. Some summary results of research by the BASIN team confirmed that basins are highly complex systems and large amounts of data are necessary to study and understand such systems in sufficient detail.

A database for evaluation and interpretation of net exchanges/fluxes has been made available containing pollutant concentrations in air (from passive air samplers), atmospheric deposition, water and soil samples in the catchments of Danube, Elbe, Ebro and Brévilles.

The above findings also link to modelling activities. For instance, for preferential flow, substantial improvements have been introduced in the software S1D, developed by the Czech Technical University in Prague, and the software package MARTHE, developed by BRGM. These two models showed that at small scale transport of pollutants in macro-pores is more important than previously thought. For larger scale considerations, the COMPUTE team developed and validated process-based numerical models in order to describe surface- and subsurface flow and transport. This helps to quantify and to predict the fate of contaminants at the catchment scale. Demonstration simulations were applied in the Beerze-Reusel area (Meuse Basin) and in the Geer Catchment. In addition, the code "Geotop" introduced new hydrological models for quantification of flow and solute transport at the catchment to river basin scale with a theoretical framework based on the travel time concept. First applications to this challenge were for instance applied to stream flows in the Gallego subcatchment of the Ebro Basin. Based on these achievements new numerical simulations are now underway for the entire Ebro Basin. The COMPUTE team also took up the hypothesis of FLUX that most contaminant loads are carried by rivers during rarely occurring floods.

It is recommended to complement regular monitoring activities e.g. on a monthly basis by event-based and time-integrated sampling. Ideally, modern river basin management would use wireless sensor networks in addition to existing networks.

Coherent socio-economic analyses by the INTEGRATOR team support the decision-making in river basin management by proposing several conceptual models (combining socio-economic and bio-physical data) and by orienting the presentation of integrated information towards river basin management issues and more general stakeholder demands. Main issues at stake in the various basins were identified and analysed. These issues include (i) diffuse (agriculture) & point source pollution, (ii) contaminated sediments & floodplain soils, (iii) flooding, large scale groundwater pollution due to past mining/industry activity, (iv) damming/reservoirs & resulting sediment deficits downstream, (v) soil erosion, (vi) water scarcity, and (vii) salinisation.

The key findings of AquaTerra, validated by two rounds of consultation with WP leaders, are summarized in fact sheets of data and knowledge generated, as well as of the tools produced in AquaTerra. Stakeholders workshops organised by INTEGRATOR, BASIN and EUPOL and associated results were used to compile the stakeholders' demands related to the overall issue of river basin management.

To make AquaTerra's outcomes accessible to a wide range of end users, research results are presented via the AquaTerra On-line Information System (ATOIS) web portal.

A key recommendation by the socioeconomic work conducted in AquaTerra is that communication between information providers and stakeholders is vitally important in order to evaluate the range of methods and approaches to solve them. To complement this work, the EUPOL team identified 54 policy questions in brainstorming sessions, questionnaires and interviews. These were grouped into broad categories of reference, to show which scientific areas were of particular interest to policy makers. The questions were related to land use, river basin management, water and climate change, while questions related to agriculture and soil were the least common. However, European soil policy is less developed than water policy at present and climate change and river basin management are currently important topics of discussion. This led to suggestions of further interdisciplinary research methods, involvement of stakeholders in the research process, with emphasis on learning processes in policy making.